Light Scattering by Optically Soft Particles
Theory and Applications

Subodh K. Sharma and
David J. Somerford

Light Scattering by Optically Soft Particles

Theory and Applications

Published in association with
Praxis Publishing
Chichester, UK

Dr Subodh K. Sharma
S. N. Bose National Centre for Basic Sciences
Salt Lake
Kolkata
India

Dr David J. Somerford *(deceased)*
Formerly of Department of Physics and Astronomy
University of Wales
College of Cardiff
Cardiff
UK

SPRINGER–PRAXIS BOOKS IN ENVIRONMENTAL SCIENCES *(LIGHT SCATTERING SUB-SERIES)*
SUBJECT *ADVISORY EDITOR*: John Mason B.Sc., M.Sc., Ph.D.
EDITORIAL *ADVISORY BOARD MEMBER*: Dr. Alexander A. Kokhanovsky, Ph.D. Optics, Institute of Physics, Minsk, Belarus. Currently at the Institute of Environmental Physics, University of Bremen, Bremen, Germany

---

ISBN 3-540-23910-3 Springer-Verlag Berlin Heidelberg New York

Springer is part of Springer-Science + Business Media (springeronline.com)

Bibliographic information published by Die Deutsche Bibliothek

Die Deutsche Bibliothek lists this publication in the Deutsche Nationalbibliografie; detailed bibliographic data are available from the Internet at http://dnb.ddb.de

Library of Congress Control Number: 2006925167

Apart from any fair dealing for the purposes of research or private study, or criticism or review, as permitted under the Copyright, Designs and Patents Act 1988, this publication may only be reproduced, stored or transmitted, in any form or by any means, with the prior permission in writing of the publishers, or in the case of reprographic reproduction in accordance with the terms of licences issued by the Copyright Licensing Agency. Enquiries concerning reproduction outside those terms should be sent to the publishers.

© Praxis Publishing Ltd, Chichester, UK, 2006
Printed in Germany

The use of general descriptive names, registered names, trademarks, etc. in this publication does not imply, even in the absence of a specific statement, that such names are exempt from the relevant protective laws and regulations and therefore free for general use.

Cover design: Jim Wilkie
Project management: Originator Publishing Services, Gt Yarmouth, Norfolk, UK

Printed on acid-free paper

Contents

Preface ix
List of figures xi
List of tables xiii
List of abbreviations xv

1 Introduction 1

2 The eikonal approximation in non-relativistic potential scattering 5
   2.1 Preliminaries of the problem 5
   2.2 The wave function in the eikonal approximation 6
      2.2.1 Approximation from the Schrödinger equation 6
      2.2.2 Approximation from the integral equation 7
      2.2.3 Propagator approximation 8
      2.2.4 Physical picture 9
   2.3 Scattering amplitude 9
      2.3.1 Eikonal amplitude 9
      2.3.2 Glauber variant 10
   2.4 Relationship with partial wave expansion 11
   2.5 Comparison with the Born series 12
   2.6 Interpretation as a long-range approximation 13
   2.7 Numerical comparisons 13
   2.8 Modified eikonal approximations 14
      2.8.1 The eikonal expansion 14
      2.8.2 The eikonal–Born series 15
      2.8.3 The generalized eikonal approximation 16

Contents

3	Eikonal approximation in optical scattering	17
	3.1 Definitions and terminology	19
	3.2 Analogy with potential scattering	21
	3.3 Validity of scalar scattering approximation	23
	3.4 Scattering by a homogeneous sphere	24
	3.4.1 The eikonal approximation	24
	3.4.2 Derivation from the Mie solutions	31
	3.4.3 Relationship with the anomalous diffraction approximation	33
	3.4.4 Corrections to the eikonal approximation	36
	3.4.5 Numerical comparisons	41
	3.4.6 One-dimensional models	45
	3.4.7 Backscattering	47
	3.4.8 Vector description	50
	3.5 Infinitely long cylinder	51
	3.5.1 Normal incidence	51
	3.5.2 Derivation from exact solutions	57
	3.5.3 Vector formalism	59
	3.5.4 Corrections to the eikonal approximation	60
	3.5.5 Numerical comparisons	63
	3.5.6 Oblique incidence	65
	3.6 Coated spheres	66
	3.7 Spheroids and ellipsoids	70
	3.8 Some other shapes	74
	3.8.1 Columnar particles	74
	3.8.2 Cube	76
	3.8.3 Plates and needles	76
	3.8.4 Parallelepiped	77
	3.8.5 Statistical interpretation of the ADA	78
	3.9 Randomly oriented monodisperse particles	81
	3.9.1 Long-circular and elliptic cylinders	81
	3.9.2 Hexagonal columns	83
	3.9.3 Finite cylinders	85
	3.9.4 Spheroids and ellipsoids	85
	3.9.5 Disks	87
	3.10 Polydispersion of spheres	88
4	**Other soft-particle approximations**	**91**
	4.1 Rayleigh–Gans–Debye approximation	91
	4.1.1 Homogeneous sphere	92
	4.1.2 Scattering by an infinitely long cylinder	95
	4.2 Perelman approximation	96
	4.2.1 Homogeneous sphere	96
	4.2.2 Special cases	101
	4.2.3 Backscattering	103

		4.2.4	The scalar Perelman approximation	103
		4.2.5	Infinitely long cylinder	105
	4.3	Hart and Montroll approximation	107
		4.3.1	Homogeneous sphere	107
		4.3.2	Infinitely long cylinders	109
	4.4	Evans and Fournier approximation	110
		4.4.1	Homogeneous sphere	111
		4.4.2	Homogeneous spheroids	112
	4.5	Bohren and Nevitt approximation	114
	4.6	Nussenzveig and Wiscombe approximation	117
	4.7	Penndorf–Shifrin–Punina approximation	118
	4.8	Numerical comparisons	118

5 Applications of eikonal-type approximations 127
  5.1 Particle size determination 127
      5.1.1 One particle at a time 127
      5.1.2 Suspension of particles 135
      5.1.3 Aggregates 145
  5.2 Interstellar and interplanetary dust 146
  5.3 Plasma density profiling 148
  5.4 Biomedical optics 150
      5.4.1 Blood optics 150
      5.4.2 Tissue optics 153
      5.4.3 Size and shape of bacteria 154
      5.4.4 Circular dichroism and optical rotation 154
  5.5 Ocean optics 156
  5.6 Miscellaneous applications 157

Appendices
  A Scattering formulas in the anomalous diffraction approximation
      for an arbitrarily oriented hexagonal column 159
  B Addition theorems employed in deriving the main form of the
      Perelman approximation (MPA) for a spherical particle 163
  C Derivation of Perelman approximation for the light scattered by
      an infinitely long cylinder 167
  D Mean value theorem and estimation of the key parameters of the
      distribution 171
      D.1 Estimation of the particle number N 173
      D.2 Estimation of the first moment of $f(a)$ 174
  E Pearson method 175

References 177
Index 193

To my parents

Preface

This book is devoted essentially to the treatment of approximation methods in problems of light scattering and absorption by optically soft particles (whose refractive indices are close to that of the surrounding medium). The study of approximation methods is important for a variety of reasons. The most important being that approximate theories enable deeper insight into understanding the underlying physical processes involved. It is therefore desirable that light scattering workers be familiar with classical light scattering approximations. At present, no book exists which is devoted exclusively to such approximate methods. This book is an attempt to fulfill this need. We have aimed to cover all approximations that have been used in connection with light scattered or absorbed by optically soft particles. Theory as well as applications have been presented. Soft scatterers occur in various branches of science, engineering and medicine, and thus form an important class of scatterers by themselves. I believe that this book will be a useful addition to the already existing books on light scattering and absorption.

This book was planned by me and the late Dr. D. J. Somerford quite some time ago. Unfortunately, the untimely death of David in October 2003 stalled the work for a brief period. The first draft of the book was almost complete at that time. It was left to me to update it, bring it to its final shape and complete the modalities of publication. At this point I would like to record my gratitude to David for the help I received from him when I was at University College, Cardiff. I cherish many fond memories of that time.

Many friends and colleagues have contributed to the preparation of this book and I would like to thank them all. I am greatly indebted to Dr. Alexander Kokhanovsky, who read the manuscript of the book and suggested a number of improvements. I am also grateful to Professor Alan Jones of Imperial College and Professor Binayak Dutta-Roy of the S. N. Bose National Centre for Basic Sciences for many useful suggestions. It was a pleasure to work with the publishers and printers of this book. I would like to thank them for all the help provided by them. All the figures appearing in this book were re-done by them. Finally, I thank my wife Shibani and son Kunal for being the constant source of encouragement during this work.

S. K. Sharma
Kolkata, India
May, 2006

Figures

2.1	Scattering of a plane wave by a potential in the EA	9
3.1	Scattering geometry for a homogeneous sphere	25
3.2	Typical variations in extinction efficiency curve with size parameter and refractive index	28
3.3	Scattered intensity $i(\theta)$ as a function of θ for a perfect homogeneous sphere with $n = 1.10$ and $x = 10.0$	44
3.4	Description of backscattering process of a wave by (a) single hard scattering, (b) double-scattering and (c) triple-scattering	47
3.5	(a) Comparison of $\log i(\pi)_{SS}$ with $i(\pi)_{MIE}$ for $m = 1.05$. (b) Comparison of $\log \|S(\pi)_{SS} + S(\pi)_{SS}^I\|^2$ with $i(\pi)_{MIE}$ for $m = 1.05$	48
3.6	Scattering geometry for an infinitely long cylinder	52
3.7	(a) Comparison of $i(\theta)_{TMWS}$ with $i(\theta)$ corresponding to (3.98a). The refractive index is $m = 1.05$ and $x = 10.0$. (b) Comparison of $i(\theta)_{TEWS}$ with $i(\theta)$ corresponding to (3.98b). The refractive index $m = 1.05$ and $x = 10.0$	61
4.1	Percent error in extinction efficiency *versus* x for MPA for $m = 1.05$, $m = 1.10$ and $m = 1.15$	120
4.2	Comparison of variation of $ln[I(\theta)]$ with θ in the MPA with the corresponding exact result for $m = 1.05$ and $x = 20.0$	120
4.3	Percent error in extinction efficiencies against size parameter for $m = 1.06$	121
4.4	Contour plot over the complex index of the refraction plane for extinction efficiency. Maximum percent error between Mie theory and the EFA	123
4.5	Absorption efficiencies of water droplets at $\lambda = 2.0\,\mu\text{m}$ calculated exactly and approximately	124
4.6	Comparison of ZKA, BNA, STA and Mie calculations	125
5.1	Scattered intensity ratio $R(\theta_1, \theta_2)$ *versus* size parameter for an infinitely long homogeneous cylinder of $m = 1.05$ and for the angle pair $(5°, 2.5°)$	129
5.2	Locations of the first minimum *versus* size parameter for $m = 1.50$ for an infinitely long homogeneous cylinder	132
5.3	Typical variations of extinction efficiency factor with size parameter and relative refractive index	138

Tables

3.1	Analytic approximations and their domains of validity	18
3.2	The ratio $i(0)_{exact}/i(0)_{scalar}$ for $n = 1.15$	23
3.3	Percent error in various approximate methods in $i(0)$ for a homogeneous sphere of refractive index 1.05	35
3.4	Percent error in various approximate methods in Q_{ext} for a homogeneous sphere of refractive index 1.05	42
3.5	Comparison of the average separation between two successive minima as predicted by (3.107)	49
3.6	Percent error in various approximate methods in $i(0)_{TMWS}$ for an infinitely long homogeneous cylinder of refractive index 1.05	63
3.7	Percent error in various approximate methods in $i(0)_{TEWS}$ for an infinitely long homogeneous cylinder of refractive index 1.05	64
4.1	The percent errors in $i(0)_{RGDA}$ for scattering of unpolarized incident radiation by a homogeneous sphere of $m = 1.05$	119
4.2	The percent errors in $i(0)_{RGDA}$ for $m = 1.05$ for a homogeneous, infinitely long cylinder at perpendicular incidence	119
4.3	The ρ values below which the extinction in the MPA gives errors that are $<5\%$	119
4.4	Percent error in various approximate methods in $1.0 \leq x \leq 10.0$ for Q_{ext} for a homogeneous sphere. The relative refractive index is $m = 1.06$	121
4.5	Percent error in various approximate methods in $1.0 \leq x \leq 10.0$ for $i(0)$ for a homogeneous sphere. The relative refractive index is $m = 1.06$	122
4.6	Percent error in various approximations in $1.0 \leq x \leq 25.0$ for $i(0)$ for a homogeneous sphere. The relative refractive index is $m = 1.05$	123
5.1	Percent error in size parameter determination for $m = 1.05$ in using FCI	129
5.2	Percent error in size parameter determination for $m = 1.15$ in using FCI	130
5.3	Percent error in size determination from the position of the first minimum using RGDA, RGDA1, and RGDA2 for $n = 1.05$, and percent errors in the FCI	134
5.4	Verification of formulas (5.30), (5.32), (5.36) and (5.38) for the locations and magnitudes of the first maximum and the first minimum of K_{ext}	141

Abbreviations

ADA	Anomalous Diffraction Approximation
ADT	Anomalous Diffraction Theory
BNA	Bohren and Nevitt Approximation
CD	Circular Dichroism
DDA	Discrete Dipole Approximation
DI	Deformation Index
EA	Eikonal Approximation
EADA	Extended Anomalous Diffraction Approximation
EFA	Evans and Fournier Approximation
EP	Eikonal Picture
FCI	First Order Corrected Eikonal Approximation
FCII	First Order Corrected Anomalous Diffraction Approximation
FDTD	Finite Difference Time Domain
FMA	Fymat and Mease Approximation
FPR	Fabry–Perot Resonator
GEA	Generalized Eikonal Approximation
GO	Geometrical Optics
GRA	Gaussian Ray Approximation
HMA	Hart and Montroll Approximation
IPA	Improved Perelman Approximation
KZA	Kokhanovsky and Zege Approximation
LS	Large Scatterer
MADA	Modified Anomalous Diffraction Approximation
MDF	Modified Diffraction Formula
MEP	Modified Eikonal Picture
MFCI	Modified First Order Corrected Eikonal Approximation
MGEA	Modified Generalized Eikonal Approximation
MPA	Main form of the Perelman Approximation

MSPA	Modified Scalar Perelman Approximation
NWA	Nussenzveig and Wiscombe Approximation
OR	Optical Rotation
PA	Perelman Approximation
PSPA	Penndorf–Shifrin–Punina Approximation
QSA	Quasi Static Approximation
RBC	Red Blood Cell
RGDA	Rayleigh–Gans–Debye Approximation
SADT	Aimplified Anomalous Diffraction Theory
SPA	Scalar Perelman Approximation
TE	Transverse Electric
TEM	Transverse Electric and Magnetic
TEWS	Transverse Electric Wave Scattering
TM	Transverse Magnetic
TMWS	Transverse Magnetic Wave Scattering
WKBA	Wentzel–Kramers–Brillouin Approximation

1

Introduction

Light scattered by an obstacle is related to its physical properties and, hence, in principle it is possible to obtain information about the scatterer from analysis of the scattered light. Thus, for many years, the light scattering technique has been used for inferring the size, shape and refractive index of particles in various scientific disciplines. Areas of interest include bio-particles, macromolecules, colloids, aerosols, hydrosols, and geological and astrophysical particles. The fact that it is a nondestructive technique and can be used for atmospheric and astrophysical particles, which are not easily accessible, makes it a very attractive diagnostic tool. In addition, many instruments based on this technique have been developed for routine measurements in industry.

Unfortunately, the problems involving the scattering of light by arbitrary shaped scatterers are so complex that exact solutions are not always known. But even when exact solutions are known these are usually complicated and computationally tedious, and it has been long desirable to obtain simple approximate formulas which also provide physical insight into the scattering processes. Commonly used approximations for this purpose are Rayleigh, Rayleigh–Gans–Debye, anomalous diffraction, Fraunhofer diffraction and geometrical optics. Most books on light scattering or electromagnetic scattering devote considerable attention to the discussion of these approximations.

Many particles of interest in nature are such that their refractive index m, relative to the surrounding medium is close to unity (i.e., $|m - 1| \ll 1$). For example, for a large number of biological particles m lies between 1.01 and 1.10, for many aerosols m is less than 1.5, and for particles of interest in ocean optics the relative refractive index is less than 1.15 (e.g., relative refractive index of algae with respect to water is 1.10). The particles satisfying the condition $|m - 1| \ll 1$ are generally referred to as soft particles. Because of the coincidence that soft-particle abundance in nature is very large, there is great interest in approximation methods related to light scattering and absorption by such particles. We restrict ourselves to elastic scattering in this book.

2 Introduction

That is, it is assumed that the frequency of incident light remains unaltered after the scattering process. This excludes from the ambit of this book inelastic processes like Raman and Brillouin scattering.

This book focuses on the class of approximations suitable for describing light scattering and absorption by soft particles and on the interrelationships between these approximations. The most well-known among this class of approximations is the anomalous diffraction approximation. This approximation was set forth by van de Hulst (1957) and, hence, is also known as the van de Hulst approximation. This approximation has also been referred to as the physical optics approximation by Shifrin (1988). A closely related approximation is the eikonal approximation (EA). The word *eikón* is a Greek word meaning *image*. The approximation is also known as the straight line approximation in the sense that image formation in physical optics is related to the propagation of light in a straight line. For scattering description, the EA was first used successfully in high-energy scattering problems. Therefore, it has also been referred to as the high-energy approximation. In addition to these two approximations – namely, the anomalous diffraction and the EA – the other soft-particle ($|m-1| \ll 1$) approximations considered in this book are the Perelman approximation or the S-approximation, the Hart and Montroll approximation, the Evans and Fournier approximation, the Nussenzveig and Wiscombe approximation, the Penndorf–Shifrin–Punina approximation and the Bohren and Nevitt approximation. A brief description of the Rayleigh–Gans–Debye approximation has also been included. In the development of the subject matter of this book, the EA forms the core, with reference to which all other approximations have been considered. The main motivation for this approach originates from the fact that, in comparison with the anomalous diffraction approximation, the EA is a later addition and, consequently, is less familiar to workers in the field of light scattering and thus requires more exposure. Other soft-particle approximations have not been exploited to the extent necessary to acquire central importance.

The EA was first introduced in problems connected with high-energy potential and nuclear scattering in the late 1940s and the early 1950s. The earliest development of this approximation appears to be due to Molière (1947) in his work on the elastic scattering of fast charged particles. In Soviet literature, this approximation has also been referred to as the Sitenko–Glauber approximation after the work of Sitenko (1959). An approximation similar to the EA was obtained by Raman and Nath (1935, 1936) in the context of the scattering of light by acoustic waves. The eikonal method has been generalized by Weinberg (1962) for waves with several components propagating in inhomogeneous anisotropic media. A detailed account of the early development of the EA in non-relativistic potential and nuclear scattering has been given by Glauber (1959) in his famous Boulder lectures on high-energy collision theory.

Despite its extensive use in nuclear scattering problems, it was quite some time before applications of the EA to new disciplines started appearing. It was extended to relativistic potential scattering of spin $1/2$ particles by Hunziker (1963) and Baker (1964). Franco (1968) demonstrated its usefulness in calculations of $e^- - H(1s)$ elastic scattering amplitude. Applications to electron–molecule and atom–atom scattering followed soon (Byron *et al.*, 1970; Yates and Tenny, 1972a, b; and Chang *et al.*, 1973).

At about the same time, it was used successfully to sum the high-energy behavior of generalized ladder diagrams in quantum field theory in a compact and useful manner (Abarbanel and Itzykson, 1969; Englert *et al.*, 1969; Lévy and Sucher, 1969). Drawing an analogy between a potential and a refractive medium, the EA was then adapted to scattering problems in optics. Although this suggestion first came quite early (Greenberg, 1960), there seems to have been no activity till Borovoi and Krutikov (1976) applied it to the scattering of light by an ensemble of particles and Sharma and Debi (1978) independently suggested its use in the analysis of light scattered by biological cells.

In the 1970s and later the EA has been applied to many other scattering problems. Gersten and Mittleman (1975) applied it to the scattering of charged particles in the presence of an electromagnetic field. The process is of central importance in the study of plasma heating by electromagnetic waves. In the same year it was applied to the atomic scattering from corrugated repulsive walls (Garibaldi *et al.*, 1975). It was found that with very little calculational effort it could reproduce the experimental results reasonably well. Group theoretical techniques in eikonal physics were initiated by Harnad (1975) and by León *et al.* (1977). Stepanov and Shelagin (1986) studied the scattering of very cold neutrons from heterogeneities in condensed matter while Schulp (1989) used the EA for inversion of atomic and molecular beam scattering data. Gómez and Castaño (1988) showed that the multi-slice approximation used in the context of transmission electron microscopy (Cowley and Moodie, 1957) was nothing but the EA, while the real space method (van Dyck and Coene, 1984) is in fact a modified form of the EA. The inverse scattering problem in the EA has been discussed by Varsimashvili *et al.* (1980). Sharma (1986) and Sharma and Dasgupta (1987) pointed out its potential applications in plasma diagnostics. It has also been used to investigate the elastic electron–ion collision processes in strongly coupled plasmas (Jung, 1996), and recently Sharma and Saha (2004) have applied this approximation to acoustic scattering by weak scatterers. Currently, the EA is being used extensively in the context of a variety of scattering problems in various disciplines.

Although the EA made its debut as a high-energy approximation, studies on its validity domain in various disciplines show that this need not be the case always. For example, it fails as a high-energy approximation in field theories describing the interaction of two scalar particles via a scalar meson exchange (Banerjee and Mallik, 1974). In these theories, it needs to be viewed as a long-range approximation (Banerjee *et al.*, 1977; Banerjee and Sharma, 1978). That is, the interpretation of the validity domain of the EA is dependent on the nature of the interaction involved. Thus, despite the use of the term "high-energy approximation", a more appropriate choice of name seems to be the "eikonal approximation". The term EA is at times also used to describe a class of approximations which approximate wave propagation in a medium which is slowly varying in space compared with the wavelength of the wave. It is in this sense that the semi-classical approximations are at times also referred to as the EA (Walters, 1984).

Historically, the term "eikonal" first appeared in optics when Bruns (1895) used it independently to describe some functions similar to the characteristic functions of Hamilton (1828) (see, e.g., Born and Wolf, 1970). Somerfeld and Runge (1911) showed that a convenient starting point for obtaining the eikonal equation, which characterizes the eikonal function, is the scalar wave equation of optics:

$$\nabla^2 \psi(\mathbf{r}) + m^2(\mathbf{r})k^2\psi(\mathbf{r}) = 0, \tag{1.1}$$

where $k = (2\pi/\lambda)$ is the wave number in vacuum, λ is the wavelength in vacuum and $m(\mathbf{r})$ is the refractive index of the medium. The complex amplitude ψ may be written in the form:

$$\psi(\mathbf{r}) = A \exp(ikS), \tag{1.2}$$

where $A(x, y, z)$ is the real amplitude and $kS(x, y, z)$ is the phase. The function S is called the "eikonal". Substitution of (1.2) in (1.1) leads to:

$$\frac{1}{k^2 A}\nabla^2 A + \frac{2i}{kA}\nabla A.\nabla S - (\nabla S)^2 + \frac{i}{k}\nabla^2 S + m^2 = 0. \tag{1.3}$$

In the geometrical optics limit $\lambda \to 0$, the above equation reduces to:

$$(\nabla S)^2 = m^2. \tag{1.4}$$

Equation (1.4) is known as the "eikonal equation" and determines the wave propagation in the geometrical optics approximation. The surfaces $S = $ constant are the wavefronts and the normals to these surfaces represent ray directions.

The solution (1.2) for the problem of scattering of light by an object of relative refractive index $m(\mathbf{r})$ may then be written as:

$$A \exp(ikS(\mathbf{r})) = A \exp\left[i\mathbf{k}.\mathbf{r} + ik\int_{\mathbf{r}_0}^{\mathbf{r}} [m(\mathbf{r}') - 1]\, ds'\right]. \tag{1.5}$$

In (1.5) $\mathbf{r}_0$ is an arbitrary point outside the region of interaction and the integration is along the actual ray path. The first term on the right-hand side is the incident plane wave and the second term gives the change in the phase of the incident wave due to the presence of the scatterer. A particular case of this approximation is obtained if the curvature of ray paths is neglected and if A is assumed to be unity. The approximate $\psi(\mathbf{r})$:

$$\psi(\mathbf{r}) = \exp\left[ikz + ik\int_{z_0}^{z} [m(\mathbf{r}') - 1]\, dz'\right], \tag{1.6}$$

is then nothing more than what has been termed the "WKBA interior wave function" by Saxon (1955). This is also the interior wave function in the anomalous diffraction approximation (ADA) (van de Hulst, 1957). In writing (1.6), the z-axis has been chosen along the incident momentum direction. The assumption that A is unity holds only if the relative refractive index of the scatterer is close to unity. Under these circumstances one may also cast (1.6) in the form:

$$\psi(\mathbf{r}) = \exp\left[ikz + \frac{ik}{2}\int_{z_0}^{z} [m^2(\mathbf{r}') - 1]\, dz'\right]. \tag{1.7}$$

It is the solution (1.7) of (1.1) which has come to be known as the EA in optical scattering and absorption problems.

2

The eikonal approximation in non-relativistic potential scattering

Much of the work on development of the theory of the eikonal approximation (EA) for the elastic scattering of light by a dielectric object has been adapted from the work on quantum mechanical potential scattering by drawing an analogy between the potential and the refractive index of the scattering object. It is, therefore, desirable to first introduce the work on the EA in relation to potential scattering. This chapter is devoted to this task. Some of the books and articles wherein the EA has been reviewed in the context of potential scattering include Glauber (1959), Schiff (1968), Abarbanel (1972), Gerjuoy and Thomas (1974), Joachain and Quigg (1974), Joachain (1975), Byron and Joachain (1977), and Gien (1988).

2.1 PRELIMINARIES OF THE PROBLEM

We begin by considering the non-relativistic elastic scattering of a spinless particle of mass M by a local potential $V(\mathbf{r})$ of range a. Let $\mathbf{k}_i$ and $\mathbf{k}_f$ be the initial and final wave vectors associated with the particle and let θ be the scattering angle between them. The particle energy E is $\hbar^2 k^2/2M$ and $|\mathbf{k}_i| = |\mathbf{k}_f| = k$. The problem to be considered is the solution of the Schrödinger equation:

$$\left[\frac{-\hbar^2}{2M}\nabla^2 + V(\mathbf{r})\right]\psi(\mathbf{r}) = E\psi(\mathbf{r}), \tag{2.1}$$

or, alternatively, the solution of the integral equation:

$$\psi(\mathbf{r}) = \exp(i\mathbf{k}_i.\mathbf{r}) - \int G(\mathbf{r} - \mathbf{r}')U(\mathbf{r}')\psi(\mathbf{r}')\,d\mathbf{r}', \tag{2.2}$$

where
$$G(\mathbf{r} - \mathbf{r}') = \frac{\exp(ik|\mathbf{r} - \mathbf{r}'|)}{4\pi|\mathbf{r} - \mathbf{r}'|}, \qquad (2.3)$$

is the free particle Green's function. The scattering amplitude $f(\theta)$ is then given by the expression:

$$f(\theta, \varphi) = -\frac{1}{4\pi} \int e^{-i\mathbf{k}_f \cdot \mathbf{r}} U(\mathbf{r}) \psi(\mathbf{r})\, d\mathbf{r}, \qquad (2.4)$$

where $U(\mathbf{r}) = (2M/\hbar^2) V(\mathbf{r})$ is the reduced potential. Hence, if $\psi(\mathbf{r})$ is known inside the region $U(\mathbf{r}) \neq 0$, the scattering amplitude can be computed using (2.4). The basic problem thus reduces to finding the wave function inside the scattering region.

2.2 THE WAVE FUNCTION IN THE EIKONAL APPROXIMATION

The EA in scattering problems may be introduced in a number of ways. Each derivation brings out a particular facet of the approximation more clearly. A detailed account of the basic EA in the context of potential and nuclear scattering can be found in the famous Boulder lectures of Glauber (1959).

2.2.1 Approximation from the Schrödinger equation

Consider a trial solution of equation (2.1) in the form:

$$\psi(\mathbf{r}) = e^{i\mathbf{k}_i \cdot \mathbf{r}} \phi(\mathbf{r}). \qquad (2.5)$$

This, when substituted in (2.1), gives:

$$\left[\nabla^2 + 2ik \frac{\partial}{\partial z}\right] \phi(\mathbf{r}) = U(\mathbf{r})\phi(\mathbf{r}). \qquad (2.6)$$

In arriving at (2.6), the z-axis has been chosen along the direction of the incident momentum $\mathbf{k}_i$. Neglecting the term ∇^2 in the left-hand side of (2.6), we obtain:

$$2ik \frac{\partial}{\partial z} \phi(\mathbf{r}) = U(\mathbf{r})\phi(\mathbf{r}), \qquad (2.7)$$

which, with the boundary condition $\phi(-\infty) = 1$, gives the solution:

$$\phi(\mathbf{r}) = e^{i\chi(\mathbf{b}, z)}, \qquad (2.8)$$

where the phase shift function $\chi(\mathbf{b}, z)$ is:

$$\chi(\mathbf{b}, z) = -\frac{1}{2k} \int_{-\infty}^{z} U(\mathbf{b}, z')\, dz', \qquad (2.9)$$

and $\mathbf{b}$ is a vector in the impact parameter plane (x, y). The solution (2.5) along with (2.8) and (2.9) is the EA in its simplest form. Notice that this solution is formally identical to (1.7). This is the reason why the approximation (2.5) was named as the EA in potential scattering.

It can be checked that the assumptions made in arriving at (2.7) – namely:

$$\nabla^2 \phi(\mathbf{r}) \ll U(\mathbf{r})\phi(\mathbf{r}), \tag{2.10}$$

and

$$\nabla^2 \phi(\mathbf{r}) \ll 2ik\, \partial \phi(\mathbf{r})/\partial z, \tag{2.11}$$

essentially mean that the two validity criteria for the EA are:

$$(i) \quad |U_0|/k^2 \ll 1, \tag{2.12}$$

and

$$(ii) \quad ka \gg 1, \tag{2.13}$$

where $|U_0|$ is the "strength" of the potential.

2.2.2 Approximation from the integral equation

Substitution of the trial wave function (2.5) in the integral equation (2.2) gives:

$$\phi(\mathbf{r}) = 1 - \frac{1}{4\pi} \int \frac{e^{ik|\mathbf{r}-\mathbf{r}'| - i\mathbf{k}_i \cdot (\mathbf{r}-\mathbf{r}')}}{|\mathbf{r}-\mathbf{r}'|} U(\mathbf{r}')\phi(\mathbf{r}')\, d\mathbf{r}'. \tag{2.14}$$

Further, defining a new variable $\mathbf{r}'' = \mathbf{r} - \mathbf{r}'$, equation (2.14) can be recast as:

$$\phi(\mathbf{r}) = 1 - \frac{1}{4\pi} \int r''\, dr'' \int e^{ikr''(1-\mu)} U(\mathbf{r} - \mathbf{r}'')\phi(\mathbf{r} - \mathbf{r}'')\, d\varphi\, d\mu, \tag{2.15}$$

with μ as the cosine of angle between the vectors $\mathbf{k}_i$ and $\mathbf{r}''$. The largest contribution to the integral in (2.15) comes when μ values are close to unity (i.e., for those values of $\mathbf{r}''$ which lie close in direction with $\mathbf{k}_i$). Otherwise, the exponential oscillates rapidly and, if the function $U\phi$ is slowly varying, the contribution of the integral becomes negligible. If it is assumed that the product $U\phi$ varies significantly over a distance d, which is such that $kd \gg 1$, the integration over μ gives:

$$\phi(\mathbf{r}) = 1 - \frac{i}{4\pi k} \int U(\mathbf{r} - \mathbf{r}'')\phi(\mathbf{r} - \mathbf{r}'')|_{\mathbf{r}''\|\mathbf{k}_i}\, dr''\, d\varphi. \tag{2.16}$$

In arriving at (2.16), terms of order $(1/kd)$ have been ignored. Since the z-axis lies in the direction of propagation, equation (2.16) can be recast in a simpler form:

$$\phi(\mathbf{r}) = 1 - \frac{i}{2k} \int_{-\infty}^{z} U(\mathbf{b}, z')\phi(\mathbf{b}, z')\, dz'. \tag{2.17}$$

Next, differentiation of (2.17) with respect to z yields (2.7) whose solution, as we have already seen, gives the EA. This derivation shows that the EA is a good approximation when $kd \gg 1$, where d is a distance in which the product $U\phi$ varies appreciably. From (2.8), it can be seen that ϕ varies appreciably in a distance $k/|U_0|$. The potential varies appreciably in a distance a. The distance d is, therefore, the smaller of the two quantities. That is, for:

$$a < \frac{k}{|U_0|}, \quad \text{or} \quad \frac{|U_0|a}{k} < 1: \quad kd \gg 1 \quad \text{implies} \quad ka \gg 1, \tag{2.18}$$

and for:

$$a > \frac{k}{|U_0|}, \quad \text{or} \quad \frac{|U_0|a}{k} > 1: \quad kd \gg 1 \quad \text{implies} \quad |U_0|/k^2 \ll 1. \tag{2.19}$$

In either case, both the conditions, $ka \gg 1$ and $|U_0|/k^2 \ll 1$, need to be satisfied. It may be noted that no restriction has been placed on the parameter $|U_0|a/k$. In contrast, the Born approximation requires this parameter to be much less than unity and the Wentzel–Kramers–Brillouin approximation (WKBA) requires this product to be much greater than unity. The EA is, therefore, particularly useful for intermediate values of the parameter $|U_0|a/k$.

2.2.3 Propagator approximation

At high energies (i.e., $ka \gg 1$ and $|U_0|/k^2 \ll 1$), it is reasonable to assume that the particle propagates nearly undeflected from its initial direction. Consequently, the only values of the intermediate momentum, $\mathbf{p}$, in Green's function:

$$G(\mathbf{r} - \mathbf{r}') = -\frac{1}{(2\pi)^3} \int d\mathbf{p} \frac{e^{i\mathbf{p}\cdot(\mathbf{r}-\mathbf{r}')}}{p^2 - k^2 - i\epsilon}, \tag{2.20}$$

which are significant are those which are near $\mathbf{k}_i$. Introducing a new variable $\mathbf{Q} = \mathbf{p} - \mathbf{k}_i$, the propagator (2.20) can be expressed as:

$$G(\mathbf{r} - \mathbf{r}') = -\frac{e^{i\mathbf{k}_i\cdot(\mathbf{r}-\mathbf{r}')}}{(2\pi)^3} \int \frac{e^{i\mathbf{Q}\cdot(\mathbf{r}-\mathbf{r}')}}{2\mathbf{k}_i\cdot\mathbf{Q} + Q^2 - i\epsilon} d\mathbf{Q}.$$

Since the main contribution comes from values $|\mathbf{Q}| \ll k$, the denominator of the propagator can be linearized by neglecting the Q^2 term. Using the relations:

$$\frac{1}{2\pi} \int_{-\infty}^{\infty} \frac{e^{iQ_z(z-z')}}{Q_z - i\epsilon} dQ_z = \Theta(z - z'), \tag{2.21}$$

and

$$\frac{1}{4\pi^2} \int_{-\infty}^{\infty} e^{i\mathbf{Q}_\perp\cdot(\mathbf{b}-\mathbf{b}')} d\mathbf{Q}_\perp = \delta(\mathbf{b} - \mathbf{b}'), \tag{2.22}$$

the linearized (or eikonalized) propagator G_{EA}, after performing straightforward integrations, becomes:

$$G(\mathbf{r} - \mathbf{r}')_{EA} = \frac{-i}{2k} e^{ik(z-z')} \delta(\mathbf{b} - \mathbf{b}') \Theta(z - z'). \tag{2.23}$$

Here, Θ is the step function and $[d/dx]\Theta(x) = \delta(x)$ is the Dirac delta function. Employing the trial wave function (2.5) and the linearized propagator (2.23), the integral equation (2.2) can be solved without further approximation leading to the solution (2.8). The eikonal propagator (2.23) clearly exhibits straight line propagation in the forward direction. This is the reason why the EA has also been termed the "straight line approximation".

2.2.4 Physical picture

The physical picture of the scattering process which emerges from the eikonal wave function:

$$\psi(\mathbf{b}, z)_{EA} = \exp\left[ikz - \frac{i}{2k}\int_{-\infty}^{z} U(\mathbf{b}, z')\, dz'\right], \qquad (2.24)$$

may be viewed as follows. A high-energy particle is assumed to pass through the scattering potential at an impact parameter **b** in a straight line trajectory (Figure 2.1). The presence of the potential introduces a change in the phase of the incident particle wave function. Its amplitude and direction of propagation remain unaffected because of the nature of the potential. The change in the phase of the wave function of the incident particle is a linear functional of the potential. Consequently, the phases in the EA are additive. This property of additivity has allowed extensive use of the EA for scattering by composite systems.

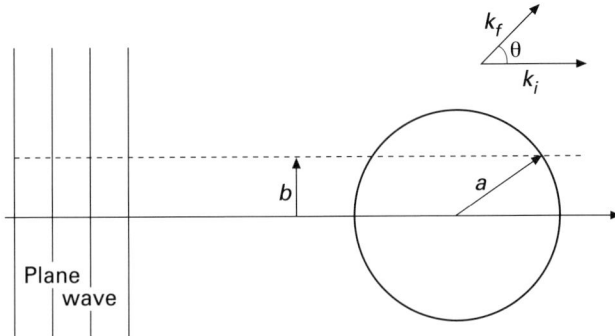

Figure 2.1. Scattering of a plane wave by a potential in the EA.

It is worth mentioning here that (2.24) does not contain an asymptotic outgoing spherical wave. Clearly, it cannot be an acceptable solution in the whole space. But, this is an acceptable solution in the range where the potential is nonzero. In fact, this is all that is needed for calculating the scattering amplitude from the formula (2.4).

2.3 SCATTERING AMPLITUDE

2.3.1 Eikonal amplitude

Substituting the approximate wave function ψ_{EA} into (2.4), the eikonal amplitude, $f(\theta)_{EA}$, may be written as:

$$f(\theta, \varphi)_{EA} = \frac{-1}{4\pi}\int e^{i(\mathbf{q}_\perp \cdot \mathbf{b} + 2kz\sin^2(\theta/2))} U(\mathbf{b}, z) e^{(-i/2k)\int_{-\infty}^{z} U(\mathbf{b}, z')dz'}\, d\mathbf{b}\, dz, \qquad (2.25)$$

where $\mathbf{q} = \mathbf{k}_i - \mathbf{k}_f$. For small-angle scattering one may approximate $\exp[2ikz\sin^2(\theta/2)]$ by unity. With this small-angle approximation, $f(\theta)_{EA}$ becomes:

$$f(\theta,\varphi)_{EA} = -\frac{1}{4\pi}\int e^{i\mathbf{q}\cdot\mathbf{b}}U(\mathbf{b},z)e^{(-i/2k)\int_{-\infty}^{z}U(\mathbf{b},z')dz'}\,d\mathbf{b}\,dz. \quad (2.26)$$

The z-integration in the above equation can be performed easily leading to a two-dimensional impact parameter representation of the scattering amplitude:

$$f(\theta,\varphi)_{EA} = -\frac{ik}{2\pi}\int e^{i\mathbf{q}_\perp\cdot\mathbf{b}}[e^{i\chi_0(\mathbf{b})} - 1]\,d\mathbf{b}, \quad (2.27)$$

where $\mathbf{b}$ is the impact parameter vector and the phase function $\chi_0(\mathbf{b})$ is:

$$\chi_0(\mathbf{b})_{EA} = -\frac{1}{2k}\int_{-\infty}^{\infty}U(\mathbf{b},z')\,dz'. \quad (2.28)$$

The amplitude (2.27) is referred to as the "eikonal amplitude". Strictly speaking though, it is a combination of the EA and an additional small-angle approximation.

For a potential which possesses azimuthal symmetry, the scattering amplitude reduces to a one-dimensional integral:

$$f(\theta)_{EA} = -ik\int J_0(kb\sin\theta)[e^{i\chi_0(b)} - 1]b\,db, \quad (2.29)$$

where J_0 is the Bessel function of order 0.

The angular range of the EA is governed by the relations $\theta^2 kd \gg 1$, which means:

$$\theta \ll \frac{1}{(ka)^{1/2}} \quad \text{if} \quad \frac{|U_0|a}{k} < 1, \quad (2.30)$$

and

$$\theta \ll \frac{|U_0|^{1/2}}{k} \quad \text{if} \quad \frac{|U_0|a}{k} > 1. \quad (2.31)$$

These inequalities follow directly from equations (2.18) and (2.19).

2.3.2 Glauber variant

In the above discussion the z-axis was chosen along the direction of incident momentum $\mathbf{k}_i$. If the z-axis is chosen along the direction $\mathbf{k}_n = (\mathbf{k}_i + \mathbf{k}_f)/2$, the momentum transfer vector $\mathbf{q} = \mathbf{k}_i - \mathbf{k}_f$ lies entirely in the impact parameter plane. Thus, $\mathbf{q}\cdot\mathbf{k}_n = 0$. The scattering amplitude (2.26) then becomes:

$$f(\theta,\varphi)_{EA} = -\frac{ik}{2\pi}\int_0^\infty e^{i\mathbf{q}_\perp\cdot\mathbf{b}}[e^{i\chi_0(\mathbf{b})} - 1]\,d\mathbf{b}, \quad (2.32)$$

even without invoking the small-angle approximation. This choice of z-direction also makes the scattering amplitude time reversal symmetric. For azimuthally symmetric potentials the angular integration gives 2π and we obtain:

$$f(\theta)_{EA} = -ik\int_0^\infty J_0(2kb\sin\theta/2)[e^{\chi_0(b)} - 1]b\,db. \quad (2.33)$$

The above choice of z-axis was suggested by Glauber (1959). Therefore, the resulting amplitude is known as the Glauber amplitude. The only difference in this amplitude from (2.27) is replacement of $kb\sin\theta$ by $2kb\sin(\theta/2)$. In what follows, it is this amplitude which will be referred to as the EA unless stated otherwise.

At first sight it may appear that by choosing the z-axis along the average momentum direction the Glauber formula has been able to avoid the small-angle approximation. But, actually, it is not so. If the z-axis is chosen along the average momentum direction right from the beginning, it can easily be shown that the scattering amplitude obtained is:

$$f_{AI}(\theta) = -ik_n \int_0^\infty J_0(2kb\sin(\theta/2))[e^{i\chi_0(b)/\cos(\theta/2)} - 1]b\,db, \qquad (2.34)$$

where $k_n = k\cos(\theta/2)$. The subscript AI stands for Abarbanel and Itzykson (1969). Numerically, however, this apparently more systematic approximation turns out to be inferior to the Glauber eikonal amplitude given in (2.33), suggesting that the terms neglected in making the small-angle approximation in fact correct certain approximations made in arriving at (2.24). That this indeed is the case has been verified by Wallace (1971, 1973a), Swift (1974), Weiss (1974) and by Banerjee *et al.* (1975). In an interesting derivation, Williams (1988) has formulated the EA utilizing the Feynman path integral method. By requiring conservation of energy, the direction of travel the particle takes in the EA has been shown to be uniquely determined to be that of the average momentum.

2.4 RELATIONSHIP WITH PARTIAL WAVE EXPANSION

If the scattering potential is a central potential, the scattering amplitude can be expressed as the following partial wave sum (see, e.g., Schiff, 1968):

$$f(\theta) = \frac{1}{2ik} \sum_{l=0}^\infty (2l+1)[e^{2i\delta_l} - 1]P_l(\cos\theta), \qquad (2.35)$$

where $P_l(\cos\theta)$ is the lth Legendre polynomial and δ_l is the phase shift suffered by the lth partial wave. The eikonalization is implemented by simply making the transformations:

$$l + \tfrac{1}{2} \to kb, \quad \sum_l \to k\int db, \quad 2\delta_l \to \chi_0(b), \qquad (2.36)$$

and

$$P_l(\cos\theta) \to J_0[2(l+\tfrac{1}{2})\sin(\theta/2)]. \qquad (2.37)$$

It can be easily verified that these transformations are valid in the high-energy limit. The validity of these transformations has been critically examined by Wallace (1973b), who showed that the partial wave sum can be converted, without any approximation, to the integral:

$$f(\theta) = -ik \int_0^\infty J_0(2kb\sin(\theta/2))[S_F(b) - 1]b\, db, \tag{2.38}$$

where

$$S_F(b) = e^{2i\delta(b)} W(\delta), \tag{2.39}$$

and $W(\delta)$ is an expansion in powers of k^{-2} with unity as its leading-order term and $\delta(b)$ as the phase shift function. In the high-energy limit ($k \to \infty$), $\delta(b)$ can be approximated by its Born approximation value:

$$\delta(b) \simeq -\frac{1}{2k} \int_0^\infty dz\, U(b, z), \tag{2.40}$$

and $W(\delta)$ may be replaced by unity. With this high-energy approximation the amplitude in (2.38) is nothing more than the eikonal amplitude.

2.5 COMPARISON WITH THE BORN SERIES

The Born series is an infinite series in powers of the interaction potential and may be expressed as:

$$f_B(\mathbf{k}_i, \mathbf{k}_f) = \sum_{n=1}^{n=\infty} f_{Bn}(\mathbf{k}_i, \mathbf{k}_f), \tag{2.41}$$

where $f_{Bn}(\mathbf{k}_i, \mathbf{k}_f)$ is the nth-order Born term. The first term in the series:

$$f_{B1} = \frac{1}{4\pi} \int d^3r\, e^{i\mathbf{q}\cdot\mathbf{r}} U(\mathbf{r}), \tag{2.42}$$

is the Born approximation. To compare the eikonal approximation with the Born series it is convenient to define an analogous eikonal series by expanding (2.32):

$$f_{EA} = \sum_{n=1}^{n=\infty} f_{EAn}(\mathbf{k}_i, \mathbf{k}_f) = \frac{k}{2i\pi} \sum_{n=1}^{n=\infty} \frac{i^n}{n!} \int e^{i\mathbf{q}\cdot\mathbf{b}} (\chi_0(b))^n\, d\mathbf{b}. \tag{2.43}$$

A comparison of (2.41) and (2.43) shows that $f_{EA1} = f_{B1}$ for all interaction potentials, energies and momentum transfers (Byron and Joachain, 1977 and references therein). But such a simple relationship does not exist among higher order terms. Nevertheless, a detailed analysis of the series for up to $n = 3$ for Yukawa-type potentials (Byron and Joachain, 1973) further suggests that the relations:

$$\lim_{k\to\infty} \frac{\operatorname{Re} f_{Bn}}{\operatorname{Re} f_{EAn}} = 1 \quad (n = \text{odd}), \tag{2.44}$$

and

$$\lim_{k\to\infty} \frac{\operatorname{Im} f_{Bn}}{\operatorname{Im} f_{EAn}} = 1 \quad (n = \text{even}), \tag{2.45}$$

hold for all n and all momentum transfers (see also Section 2.8.3). For other potentials – such as Gaussian, square well, or the polarization potential – the above relationships hold only for small-angle scattering.

Important studies in this connection have been made by Moore (1970), Weiss (1974) and Swift (1974) who examined each term of the Born series in the limit $k \to \infty$ and q fixed, and showed that when the resulting series is summed, the EA is the unique result.

2.6 INTERPRETATION AS A LONG-RANGE APPROXIMATION

High incident momentum is one way of realizing the basic assumption $ka \gg 1$. The same condition may be realized if the interaction is long-range. Indeed, for a Coulomb potential:

$$V(r) = \frac{Ze^2}{r} F(r), \qquad (2.46)$$

the EA gives exact scattered intensity. The screening $F(r)$ is necessary because the phase function $\chi_0(b)$ diverges otherwise. Subsequently, the screen is removed so that it is far from the scattering centre. If the screen $F(r)$ is chosen as a step function, $F(r) = 1$ for $r < a$ and $F(r) = 0$ for $r > a$, the scattering amplitude in the EA can be shown to be (Glauber, 1959):

$$f(\theta)_{EA} = \frac{-Ze^2/M\hbar k}{2k \sin^2(\theta/2)} e^{-2i(Ze^2/M\hbar k)\ln(2ka\sin(\theta/2))+2i\eta}, \qquad (2.47)$$

for $\theta \gg 1/ka$. In (2.47), $\eta = \arg\Gamma(1 + i(Ze^2/M\hbar k))$. The resulting scattered intensity $|f(\theta)_{EA}|^2$ is nothing more than the Rutherford formula. That is, the scattered intensity is exact in the EA for the Coulomb potential. In other words, this shows that the EA improves for long-range potentials.

2.7 NUMERICAL COMPARISONS

In the above we have considered two variants of the EA. One in which the phase accumulated by an incident wave assumes the direction of propagation to be along the direction $\mathbf{k}_i$ and the other in which the direction of propagation is assumed to be along the average momentum direction. Since the incident particle can be imagined to propagate through the interaction region in an infinite number of ways, many variants of the EA have been developed along these lines. Notable among them are those by Schiff (1956), Saxon and Schiff (1957), Abarbanel and Itzykson (1969), Sugar and Blankenbecler (1969), Lévy and Sucher (1969). Numerical comparisons of these variants with exact results have been made by Hahn (1970), Wallace (1971, 1973a), Kujawaski (1971, 1972), Berriman and Castillejo (1973), Byron *et al.* (1973, 1975), and Weiss (1974) among others. The Yukawa, exponential, square well, Wood–Saxon,

Gaussian and polarization potentials have been considered in these comparisons. The following features emerge from these comparisons:

1. When the conditions $|U_0|/k^2 \ll 1$ and $ka \gg 1$ are satisfied, the EA compares very well with the exact results. The angular domain is generally found to be larger than that expected from theoretical considerations. For certain potentials, such as Yukawa and exponential, the EA holds even at large angles. For these potentials, results are reasonably good even if $|U_0|/k^2$ is not much less than unity and ka not much greater than unity. At small angles the EA appears to work well even if $|U_0|/k^2 \gg 1$, provided the condition $ka \gg 1$ is satisfied (Byron et al., 1973).
2. The validity of the EA has been noted to be potential-dependent. But, it is not clear how to delineate the class of potentials for which the EA is valid at large angles. Nevertheless, a detailed study of the EA amplitude (Swift, 1974) and corrections to it in the limit $k \to \infty$, $q \to \infty$ for Yukawa, Gaussian and polarization potentials suggests that the EA amplitude is a good approximation to exact amplitude at all angles when $\chi_0(b)$ has a singularity at $b = 0$. If $\chi_0(b)$ is analytic at $b = 0$, the correction term dominates at large q.
3. The potential dependence of the EA has also been examined by Sugar and Blankenbecler (1969). It is concluded that the EA is a good approximation at high energy and large momentum transfers, if the potential is such that its Fourier transform does not fall off too rapidly for large momentum transfers. For example, the EA will certainly not reflect the large momentum transfer behavior for the case of a Gaussian potential. However, for potentials such as Yukawa or exponential, which fall off like a power in momentum space, the EA is expected to be an accurate approximation for a wide range of momentum transfers.

2.8 MODIFIED EIKONAL APPROXIMATIONS

2.8.1 The eikonal expansion

An eikonal expansion in powers of k^{-1} was developed by Wallace (1971, 1973a) for infinite often differentiable potentials. In this expansion the EA appears as the zeroth-order term. Corrected to first two orders, this expansion for a spherically symmetric potential may be written as:

$$f(q) = -ik \int J_0(2kb \sin \theta/2)[f(b)_{EA}^{I,II} - 1], \qquad (2.48)$$

where the first- and second-order corrected functions are given by:

$$f(b)_{EA}^{I} = e^{i(\chi_0(b) + \tau_1(b))}, \qquad (2.49)$$

and

$$f(b)_{EA}^{II} = f(b)_{EA}^{I} e^{i\tau_2(b) - \omega_2(b)}, \qquad (2.50)$$

with

$$\tau_1(b) = -\frac{1}{k^3}(1+\beta_1)\int_0^\infty dz\, U^2(r), \qquad (2.51)$$

$$\omega_2(b) = b\chi_0'(b)\nabla^2\chi_0(b)/8k^2, \qquad (2.52)$$

$$\tau_2(b) = -\frac{1}{k^5}\left(1+\frac{5}{3}\beta_1+\frac{1}{3}\beta_2\right)\int_0^\infty U^3(r)\,dz - [b(\chi_0'(b))^3/24k^2], \qquad (2.53)$$

and

$$\beta_n = b^n\frac{\partial^n}{\partial b^n}. \qquad (2.54)$$

The prime designates "derivative with respect to b". Third-order corrections have also been obtained by Wallace (1971, 1973a) but are not given here. Identical corrections have been re-derived by Baker (1972, 1973), Wallace (1973b), Swift (1974), Weiss (1974) employing alternative methods. Fourth-order corrections have been obtained by Sarkar (1980). It can easily be checked that all phase-type corrections vanish for the Coulomb potential. This is as it should be, because the EA is exact for the Coulomb potential and, hence, the corrections should be 0. Numerical evidence of systematic improvement with each increasing order (up to third order) of the eikonal expansion has been presented by Wallace (1971, 1973a) and Weiss (1974) for various potentials.

2.8.2 The eikonal–Born series

The eikonal–Born series interpolates between second-order Born approximation and the EA. For Yukawa-type potentials, the exact scattering amplitude for $k \gg 1$ can be expressed as (Byron *et al.*, 1973):

$$f_{ex} = f_{B1} + \left[\frac{A}{k^2}+i\frac{B}{k}\right] + \left[\frac{C}{k^2}+i\frac{D}{k^3}\right] + \cdots$$

$$\equiv f_{B1} + f_{B2} + f_{B3} + \cdots. \qquad (2.55)$$

In contrast, the eikonal series yields:

$$f_{EA} = f_{B1} + i\frac{B}{k} + \frac{C}{k^2} + \cdots = f_{B1} + f_{E2} + f_{E3} + \cdots. \qquad (2.56)$$

In the above equations A and B are second-order in potential strength and C and D are third-order in potential strength. A comparison of (2.56) with (2.55) shows that neither $f_{B1}+f_{E2}$ nor $f_{B1}+f_{B2}$ is correct to order $1/k^2$. But, because A and B are second-order in potential strength while C and D are third-order in potential strength, $f = f_{B1}+f_{B2}$ is more accurate than the eikonal amplitude for the weak coupling case. The addition of the real part of the second Born term to the Glauber amplitude thus results in impressive improvements for Yukawa-type potentials. Two alternative amplitudes corrected in this way are (Byron and Joachain, 1977):

$$f_{EBS} = f_{B1} + f_{B2} + f_{E3}, \qquad (2.57)$$

and
$$f'_{EBS} = f_E + \text{Re} f_{B2}, \tag{2.58}$$
which are correct up to order $(1/k^2)$ and have been referred to in the literature as the eikonal–Born amplitudes.

2.8.3 The generalized eikonal approximation

An alternative approach to rectify the defect of the missing $\text{Re} f_{B2}$ term in the EA is to write a generalized linear propagator of the form (Chen, 1984):
$$G(\mathbf{r} - \mathbf{r}')_{GEA} = \frac{1}{2i\alpha} \delta^2(\mathbf{b} - \mathbf{b}') \Theta(z - z') e^{i\delta(z-z')}, \tag{2.59}$$

The propagator (2.59) is a generalization of (2.23). For $\alpha = \delta = k$ this reduces to the original eikonal propagator. For $\alpha = \delta = |\mathbf{k}_i + \mathbf{k}_f|/2$ and a z-axis parallel to $(\mathbf{k}_i + \mathbf{k}_f)/2$, equation (2.59) leads to a propagator that gives the Abarbanel and Itzykson amplitude (equation 2.34).

Employing (2.59), the scattering function in the generalized eikonal approximation (GEA) takes the form:
$$f(\theta, \phi)_{GEA} = f_{B1} - \frac{1}{8i\pi\alpha} \int d^3r e^{i(\mathbf{q}\cdot\mathbf{r} - \tau z - \chi(\mathbf{r}))} U(\mathbf{r}) \times \int_{-\infty}^{z} U(\mathbf{b}, z') e^{i(kz' + i\chi)\mathbf{r}'} dz', \tag{2.60}$$

where
$$\chi(\mathbf{b}, z) = \frac{-1}{2\alpha} \int_{-\infty}^{z} U(\mathbf{b}, z') dz' \tag{2.61}$$

and $\tau = \bar{\mathbf{k}} - \delta$ with $\bar{\mathbf{k}} = \mathbf{k}_i + \mathbf{k}_f/2$. The arbitrary parameters α and δ are determined in such a way that the dominant real part as well as the dominant imaginary part of the second Born amplitude are correctly reproduced. The resulting formula has been found to work very well for Yukawa and Gaussian potentials even at large angles.

With this background of the eikonal approximation in the quantum mechanical scattering problems we now proceed to consider the main theme of the book, that is the light scattering by optically soft particles.

3

Eikonal approximation in optical scattering

Light scattering experiments generally involve measurements of scattered light either from one isolated particle at a time, or measurements of scattered light from an ensemble of particles. In either case, a theory for predicting the scattering pattern from a single particle is necessary to obtain information about the scattering particle. Unfortunately, exact analytic solutions are unknown except in the simplest and most idealized cases of a sphere (Mie, 1908), an infinitely long cylinder (Rayleigh, 1881; von Ignatowsky, 1905; Wait, 1955) and a spheroid (Asano and Yamamoto, 1975). The method of Asano and Yamamoto has been improved by Farafonov (1983). A formal solution for spheroids was suggested by Moeglich (1927). Simple inhomogeneous objects such as a concentric sphere (Aden and Kerker, 1951; Güttler, 1952; Shifrin, 1952; Wait, 1963) and a concentric infinitely long cylinder (Tang, 1957; Kerker and Matijević, 1961) can also be treated exactly. For particles of other shapes one may resort to numerical procedures. Rigorous methods for spherical and non-spherical particles have been treated in many books and in recent articles (see, e.g., Barber and Hill, 1990; Ishimaru, 1997; Wriedt, 1998; Mishchenko *et al.*, 2000, 2002; and Kahnert, 2003).

Numerical solutions are generally very complex and involved. However, the complexity of numerical solutions is becoming of lesser consequence as faster and faster computers become available. Given this scenario a question sometimes asked is: Why should anyone wish to use approximations? There are many reasons for the use of approximations. First, in many applications a numerical approach still proves to be tedious or impractical or even impossible and one needs to resort to approximate methods. Second, approximation methods give simple expressions for quick and easy use and, hence, are cheap to use. Thus, whenever they offer sufficient accuracy it is desirable to use them without wasting effort on time-consuming detailed calculations. Fastness of calculations also sometimes tilts the balance in favor of approximate methods. Third, approximations provide a deeper physical and mathematical insight into the general scattering problem. An approximation may actually allow one to see what is going on during the process of interaction between radiation and the particle. This makes one aware of the important factors involved in the use of any approximation. Finally, if an approximation is sufficiently accurate it may even be used as a check on fancier numerical methods.

A survey of the early history of light scattering by a sphere can be found in Logan (1965). Simple mathematical expressions for spectral extinction and scattering properties for small-size particles have been given by Kim *et al.* (1996). A brief review of some analytic approximate methods for spherical particles has been given by Kokhanovsky and Zege (1997) and some analytic and numerical methods for elastic light scattering theories have been reviewed by Mahood (1987), Sharma and Somerford (1990, 1999), Maslowska (1991), Wriedt (1998); Wriedt and Comeberg (1998); and Jones (1999).

Over the years, a number of approximations have been designed and developed for the description of light scattering by *soft particles*. That is, for particles whose refractive index is close to that of the surrounding medium. This chapter is solely devoted to the theory of one such approximation, and some modifications/variants of this approximation including the anomalous diffraction approximation (ADA). This approximation is the eikonal approximation (EA). Other soft-particle approximations will be considered in Chapter 4. Table 3.1 shows some of the approximate methods that are employed in light scattering problems. Their validity domains are also shown alongside.

The EA has proved to be a very useful approximation for the analysis of the near forward light scattering patterns of various types of soft particles. In many cases it leads to simple analytic expressions for scattering quantities. It has been applied to the

Table 3.1. Analytic approximations and their domains of validity.

Approximation	Domain of validity
Rayleigh	$x \ll 1$; $|mx| \ll 1$
RGDA	$|m-1| \ll 1$; $2x|m-1| \ll 1$
QSA (Burberg, 1956)	Generalization of RGDA
Jobst (Jobst, 1925)	$x \gg 1$ limit of RGDA
Modified RGD (Shimizu, 1983; Gordon, 1985)	$|m-1| \ll 1, \rho \leq 1$
Geometric optics	$x \gg 1$; $x|m-1| \gg 1$
WKB	Same
Two-wave WKB (Klett and Sutherland, 1992)	Same
NWA (Nussenzveig and Wiscombe, 1980)	$x \gg 1$
ADA (van de Hulst, 1957)	$|m-1| \ll 1$; $x \gg 1$
Eikonal or high energy	Same
PA (Perelman, 1978)	Same
Walstra (Walstra, 1964)	Same
Equiphase (Chen *et al.*, 2004)	Same
HMA (Hart and Montroll 1951)	$1 < m < 1.5$; $1/2 < x < 6$
EFA (Evans and Fournier, 1990)	$|m-1| < 1$, any x
Fraunhofer diffraction	$x \gg 1$; $n' \to$ large
PSPA (Penndorf, 1962; Shifrin and Punina, 1968)	Re $S(0) \gg$ Im $S(0)$

scattering of light by particles of various shapes, such as homogeneous spheres, coated spheres, infinitely long circular cylinders, hexagonal and elliptic cylinders, spheroids and rough particles. A large variety of particles occurring in nature can be modeled using these shapes.

In the context of optical scattering the EA was introduced by drawing an analogy between the Schrödinger equation and the scalar wave equation of optics. Consequently, it is expected to be applicable in situations where the scalar description of the scattering process is sufficiently accurate to allow the vector nature of light to be ignored. Attempts have been made to include the vector character of light in the analysis of the scattering process and a reasonable degree of success has been achieved. A description of these attempts will be given at appropriate places in this chapter.

Expressions for the scattering and absorption quantities in the EA are formally identical to another approximation known as the "anomalous diffraction approximation" (ADA). In addition to diffraction, this approximation also takes into account transmitted light. It was introduced in the context of optical scattering by van de Hulst (1957) and, hence, is also known as the van de Hulst approximation. The difference is that one only needs to replace $(m^2(\mathbf{r}) - 1)$ in the EA expressions by $2(m(\mathbf{r}) - 1)$ to arrive at the corresponding ADA expressions, $m(\mathbf{r})$ being the refractive index of the scatterer relative to the refractive index of the surrounding medium. The numerical accuracy of the two approximations obviously differs. In this book we do not worry whether a particular result was obtained in the context of the EA or the ADA. All formal expressions with the subscript EA are true also for the subscript ADA with the above-mentioned replacement, and *vice versa*. To this extent the ADA and the EA have been treated on the same footing in this book.

3.1 DEFINITIONS AND TERMINOLOGY

In this section we define some scattering quantities and notations to be used in this book. If I_{inc} is the incident intensity and if the total scattered power is P_{sca}, then:

$$P_{sca} = C_{sca} I_{inc}. \tag{3.1}$$

The constant of proportionality C_{sca} has the dimension of area and is referred to as the scattering cross-section. The absorption cross-section C_{abs} is defined in an identical way and the extinction cross-section C_{ext} is the sum of the scattering and the absorption cross-sections. The scattered intensity I_{sca} is related to the incident intensity I_{inc} via the relation:

$$I_{sca} = \frac{1}{k^2 r^2} |S(\theta, \varphi)|^2 I_{inc} \equiv \frac{1}{k^2 r^2} i(\theta, \varphi) I_{inc}, \tag{3.2}$$

where k is the wave number of the incident radiation and $i(\theta, \varphi)$ is the fraction of incident energy scattered into unit solid angle about a direction that makes an angle (θ, φ) with the angle of incidence. Here, r defines the distance of the point of observation. In optics, the scattering pattern is generally depicted as plots of $i(\theta, \varphi)$ *versus* θ. In this way a constant factor is eliminated and the outcome is independent of r. In this book we also refer to $i(\theta, \varphi)$ as the scattered intensity, unless stated otherwise. Sometimes, the scattering pattern is also presented as plots of $i(\theta, \varphi)/k^2$ versus θ. This quantity is denoted as:

$$C_{sca}(\theta, \varphi) = \frac{i(\theta, \varphi)}{k^2}, \tag{3.3}$$

and is known as the differential cross-section.

The scattering function $S(\theta, \varphi)$ is related to $f(\theta, \varphi)$ (defined in Section 2.1) as:

$$S(\theta, \varphi) = -ikf(\theta, \varphi). \tag{3.4}$$

When incident radiation is treated as a scalar wave, the scattered field E_s and the incident field E_i are related via $S(\theta, \varphi)$ as:

$$E_s = \frac{e^{ikr-ikz}}{ikr} S(\theta, \varphi) E_i. \tag{3.5}$$

The scattering, absorption and extinction cross-sections, when divided by the geometrical cross-section G of the scatterer, yield the respective efficiency factors or simply the efficiencies. The efficiencies are denoted by the letter Q. Thus:

$$C_{sca} = GQ_{sca}, \quad C_{abs} = GQ_{abs}, \quad C_{ext} = GQ_{ext}. \tag{3.6}$$

The extinction efficiency is related to the real part of the forward scattering function via the so-called "optical theorem" or "extinction theorem" as:

$$C_{ext} = \frac{4\pi}{k^2} \text{Re}[S(0)]. \tag{3.7}$$

It may be mentioned here that the requirement for the applicability of the extinction theorem is that the scattering function must satisfy the unitarity property.

The scattering cross-section is related to $i(\theta, \varphi)$ and $C_{sca}(\theta, \varphi)$ via the relation:

$$C_{sca} = \frac{1}{k^2} \int_0^{2\pi} \int_0^{\pi} i(\theta, \varphi) \sin\theta \, d\theta \, d\varphi, = \int_0^{2\pi} \int_0^{\pi} C_{sca}(\theta, \varphi) \sin\theta \, d\theta \, d\varphi, \tag{3.8}$$

and the relation:

$$C_{abs} = C_{ext} - C_{sca}, \tag{3.9}$$

gives the absorption cross-section.

In multiple scattering problems, instead of $i(\theta, \varphi)$, it is sometimes preferable to employ a function $p(\theta, \varphi)$ called the "phase function". Its relation with $i(\theta, \varphi)$ is given by the equation:

$$p(\theta, \varphi) = \frac{4\pi i(\theta, \varphi)}{k^2 C_{sca}}, \tag{3.10}$$

and the normalization condition is:

$$\frac{1}{4\pi} \int_0^{2\pi} \int_0^{\pi} p(\theta, \varphi) \sin\theta \, d\theta \, d\varphi = 1. \tag{3.11}$$

The phase function $p(\theta, \varphi)$ may be thought of as the probability of a photon being scattered in a direction (θ, φ) relative to the incident direction. Another quantity of interest in multiple scattering problems is albedo ω_0. This is defined as:

$$\omega_0 = \frac{Q_{sca}}{Q_{ext}} = 1 - \frac{Q_{abs}}{Q_{ext}}. \tag{3.12}$$

Thus, for a non-absorbing scatterer or for a highly scattering object ($Q_{sca} \gg Q_{abs}$), $\omega_0 = 1$. For a totally absorbing large particle $Q_{ext} = 2Q_{abs}$ and, hence, $\omega_0 = 1/2$. A related quantity is co-albedo $\bar{\omega} = 1 - \omega_0$. One also defines an asymmetry parameter g as:

$$g = \frac{1}{2} \int_0^\pi p(\theta) \sin\theta \cos\theta \, d\theta, \tag{3.13}$$

which is a measure of asymmetry of the phase function around $\theta = 90°$. The values $g = 1, 0, -1$ correspond to forward-peaked, symmetric and backward-peaked phase functions, respectively.

The asymmetry parameter $\langle \cos\theta \rangle$ and the extinction efficiency σ_{ext} of a tenuous polydispersion $f(a)$ are given by the following equations:

$$\langle \cos\theta \rangle = \frac{\int_0^\infty f(a) g C_{sca} \, da}{\int_0^\infty f(a) C_{sca} \, da}, \tag{3.14}$$

and

$$\sigma_{ext} = N \int_0^\infty f(a) C_{ext} \, da, \tag{3.15}$$

where N is the number of scatterers per unit volume. The σ_{abs} and σ_{sca} are defined in an identical fashion.

3.2 ANALOGY WITH POTENTIAL SCATTERING

Consider a scalar wave characterized by the field $\psi(\mathbf{r})$ propagating through a medium of spatially varying relative refractive index $m(\mathbf{r})$. The field $\psi(\mathbf{r})$ then satisfies the wave equation:

$$\nabla^2 \psi(\mathbf{r}) + k^2 m^2(\mathbf{r}) \psi(\mathbf{r}) = 0. \tag{3.16}$$

A comparison of equation (3.16) with the Schrödinger equation (2.1) shows that the role played by the refractive index in (3.16) is analogous to the role played by the potential in (2.1). Thus, the various scattering quantities obtained in the context of potential scattering can be used *mutatis mutandis* in scalar optical scattering by making the replacement:

$$U(\mathbf{r}) = \left[1 - m^2(\mathbf{r})\right] k^2. \tag{3.17}$$

The scattering amplitude given by (2.4) in conjunction with (3.4) then becomes:

$$S(\theta, \varphi) = \frac{ik^3}{4\pi} \int d\mathbf{r}\, e^{-i\mathbf{k}_f \cdot \mathbf{r}} [1 - m^2(\mathbf{r})] \psi(\mathbf{r}). \tag{3.18}$$

Various approximations can be obtained for the scattering function $S(\theta, \varphi)$ by invoking different approximate forms for $\psi(\mathbf{r})$.

Having identified the similarity between a scattering potential in the Schrödinger equation and a refractive index in the scalar wave equation, it can easily be seen from (2.8) that for optical scattering $\psi(\mathbf{r})$ may be approximated as:

$$\psi(\mathbf{r})_{EA} = \exp\left[i\mathbf{k}_i \cdot \mathbf{r} + \frac{k}{2} \int_{-\infty}^{z} (m^2(\mathbf{b}, z') - 1)\, dz'\right]. \tag{3.19}$$

The EA for the scattering function is then obtained by substituting $\psi(\mathbf{r})_{EA}$ into (3.18). The z integration in (3.18) can then be easily performed. The result is:

$$S(\theta, \varphi)_{EA} = -k^2 \int d\mathbf{b}\, e^{i\mathbf{q} \cdot \mathbf{b}} [e^{i\chi(\mathbf{b})_{EA}} - 1], \tag{3.20}$$

where

$$\chi(\mathbf{b})_{EA} = (k/2) \int_{-\infty}^{\infty} [m^2(\mathbf{r}) - 1]\, dz, \tag{3.21}$$

and depends on the size, shape, internal structure and refractive index profile of the scatterer. The two-dimensional integration in (3.20) is over the projected particle area of the scatterer.

The conditions for the validity of the EA in potential scattering given by (2.12) and (2.13) translate to:

$$(i) \quad |m(\mathbf{r}) - 1| \ll 1, \tag{3.22}$$

and

$$(ii) \quad x \gg 1, \tag{3.23}$$

in optical scattering. For an absorbing particle the refractive index is complex ($m(\mathbf{r}) = n(\mathbf{r}) + in'(\mathbf{r})$) and the condition (3.23) is equivalent to two conditions:

$$(a) \quad |n(\mathbf{r}) - 1| \ll 1 \tag{3.24}$$

and

$$(b) \quad |n'(\mathbf{r})| \ll 1. \tag{3.25}$$

The dimensionless parameter $x = ka$ is known as the size parameter and is essentially a measure of the size of the scatterer in units of wavelength of the scattering radiation. The notation x has also been used in the coordinate system $\mathbf{r} \equiv (x, y, z)$. However, this does not result in any confusion and has been followed in most books. The requirement (3.22) ensures that at boundaries there is no deviation of the incident ray and that the energy reflected is negligible. The second requirement ensures that the ray travels undeviated through the scatterer as the refractive index varies slowly in wavelength.

The angular range given by (2.30) and (2.31), now converts to:

$$\theta \ll \frac{1}{\sqrt{x}} \quad \text{for} \quad x|m^2 - 1| < 1, \tag{3.26}$$

and

$$\theta \ll |m^2 - 1|^{1/2} \quad \text{for} \quad x|m^2 - 1| > 1. \tag{3.27}$$

Here, m may be taken as the maximum value of the refractive index.

An alternative derivation of (3.20) has been given by van de Hulst (1957). The outline of the method is as follows. As a consequence of conditions (3.22) and (3.23), incident rays are assumed to pass through the scattering object undeviated. A ray at an impact parameter $\mathbf{b}$ then accumulates a phase shift $\chi(\mathbf{b})_{ADA} = k \int_{-\infty}^{\infty} [m(\mathbf{r}) - 1] \, dz$. The total field in the near zone behind the scatterer can be easily found. A direct application of Huygens' principle then gives:

$$S(\theta, \varphi)_{ADA} = -k^2 \int d\mathbf{b} \, e^{i\mathbf{q}\cdot\mathbf{b}} [e^{i\chi(\mathbf{b})_{ADA}} - 1], \tag{3.28}$$

where

$$\chi(\mathbf{b})_{ADA} = k \int_{-\infty}^{\infty} [m(\mathbf{r}) - 1] \, dz, \tag{3.29}$$

which is nothing more than the scattering function in the ADA. Equation (3.28) has also been derived from the integral equation by Saxon (1955).

3.3 VALIDITY OF SCALAR SCATTERING APPROXIMATION

To assess the errors introduced by the scalar approximation, Sharma et al. (1981, 1982) have examined the scalar approximation numerically for forward scattering. Table 3.2 shows the values of the ratio $i(0)_{exact}/i(0)_{scalar}$ for the scattering of unpolarized light by a homogeneous sphere and for the transverse electric wave scattering (TEWS) by an infinitely long, homogeneous, circular cylinder. It can be concluded from the results in this table that for near forward scattering the approximation of light by a scalar wave is a reasonably good approximation for $x \geq 3.0$. It is well-known that, for transverse magnetic wave scattering (TMWS), Maxwell's equations when transformed to wave equations reduce to the scalar wave

Table 3.2. The ratio $i(0)_{exact}/i(0)_{scalar}$ for $n = 1.15$.

		x				
		1	2	5	10	20
$\dfrac{i(0)_{exact}}{i(0)_{scalar}}$	Sphere	0.791	0.918	0.953	0.975	0.982
	Cylinder (TEWS)	0.769	0.890	0.956	0.976	0.970

From Sharma and Somerford (1999).

equation without any approximation. That is, there is no error in TMWS due to scalar approximation. The errors in the EA for TMWS case, therefore, are expected to be smaller. This can be seen in Tables 3.6 and 3.7 (see pp. 63 and 64, respectively) where the errors in TMWS for x close to 1 can be seen to be much smaller in comparison with errors in TEWS.

A more general study to examine the validity of the scalar approximation has been done by Arnush (1964) for the case of a spherically symmetric continuous dielectric. In this study, it is concluded that the scattered intensity may be obtained from the scalar approximation accurately if the inequality:

$$\sqrt{\varepsilon(r)}\frac{d}{dr}\frac{1}{\sqrt{\varepsilon(r)}} \ll k^2 \varepsilon(r),$$

is satisfied for all r, $\varepsilon(r)$ being the dielectric function.

3.4 SCATTERING BY A HOMOGENEOUS SPHERE

The scattering problem for a homogeneous dielectric spherical particle is exactly soluble for arbitrary values of x and m. Thus, it is not surprising that the model of light scattering by a homogeneous sphere has been extensively studied and widely used as a standard to examine the validity domains of approximation methods. In this chapter, we derive various scattering quantities within the framework of the EA and its variants, for the scattering of light by a dielectric homogeneous sphere and also examine the validity of these approximate results *vis-à-vis* exact results numerically. The exact analytic solutions for this problem were first given by Gustav Mie (1908), and, therefore, homogeneous dielectric spheres are also referred to as "Mie particles". The resulting solution is referred to as the "Mie solution".

3.4.1 The eikonal approximation

Consider a homogeneous sphere of radius a and refractive index $m(\mathbf{r}) = m$. Because of the spherically symmetric nature of the scatterer, integration over the azimuthal angle in the general expression for the scattering function (3.20) can be easily performed. This yields:

$$S(\theta)_{EA} = -k^2 \int_0^a b\, db\, J_0(2kb \sin(\theta/2))[\exp(i\chi(b)_{EA}) - 1]. \tag{3.30}$$

The distance traveled by the ray in the scattering medium at an impact parameter b is $2\sqrt{a^2 - b^2}$ (see Figure 3.1). Thus, the function $\chi(b)_{EA}$ can be written as:

$$\chi(b)_{EA} = x(m^2 - 1)\sqrt{1 - \frac{b^2}{a^2}} = \rho^*_{EA}\sqrt{1 - \frac{b^2}{a^2}}. \tag{3.31}$$

The quantity ρ^*_{EA} is:

$$\rho^*_{EA} \equiv \rho_{EA} + i\kappa_{EA}, \tag{3.32}$$

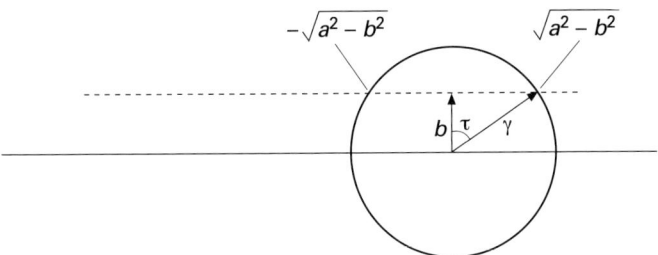

Figure 3.1. Scattering geometry for a homogeneous sphere.

where
$$\rho_{EA} = x(n^2 - n'^2 - 1), \tag{3.33}$$
and
$$\kappa_{EA} = 2xnn'. \tag{3.34}$$

The first term on the right-hand side of (3.32) is sometimes also referred to as the "phase parameter" and gives the phase shift suffered by the central ray in passing through a diameter. The quantity ρ_{EA}^* is also expressed as:
$$\rho_{EA}(1 + i\tan\beta_{EA}), \tag{3.35}$$
where
$$\rho_{EA}\tan\beta_{EA} = 2nn'x, \tag{3.36}$$
is that part of ρ_{EA}^* which describes the decay of amplitude due to absorption. Thus, the second parameter on the right-hand side of (3.32) is also referred to as the "absorption parameter".

The scattering function given by (3.30) is formally the same as the expression for $S(\theta)$ in the ADA (van de Hulst, 1957). The only difference between the two approximations is that – in place of ρ_{EA}^* – one has:
$$\rho_{ADA}^* = \rho^* = 2x(m-1) = 2x(n-1) + 2ixn',$$
$$\equiv \rho_{ADA} + i\kappa_{ADA} \equiv \rho + i\kappa. \tag{3.37}$$

Clearly, as $n \to 1$ and $n' \to 0$ the two approximations tend to the same limit. The inequalities (3.22) and (3.23) define the validity domain of this approximation too.

An alternative one-dimensional integral form often used to express $S(\theta)_{EA}$ is:
$$S(\theta)_{EA} = x^2 \int_0^{\pi/2} J_0(z\sin\gamma)\left[1 - e^{\omega^*\cos\gamma}\right]\cos\gamma\sin\gamma\,d\gamma, \tag{3.38}$$

where $z = 2\sin(\theta/2)$ and:
$$\omega^* = ix(m^2 - 1) = i\rho_{EA}^*. \tag{3.39}$$

In arriving at (3.38) from (3.30) a change of variable $b = a\sin\gamma$ has been made. While the value $\gamma = 0$ corresponds to the central ray, the value $\gamma = \pi/2$ amounts to grazing incidence.

Integration in (3.38) can be carried out analytically for forward scattering and one obtains:

$$S(0)_{EA} = x^2 \mathcal{K}(\omega^*), \qquad (3.40)$$

where

$$\mathcal{K}(\omega^*) = \frac{1}{2} - \frac{e^{\omega^*}}{\omega^*} + \frac{e^{\omega^*} - 1}{\omega^{*2}}. \qquad (3.41)$$

For non-forward scattering the integration in (3.30) or (3.38) can be performed analytically only for two special cases.

(i) For a completely absorbing sphere $n' \to \infty$, $\exp(-\rho_{EA} \tan \beta_{EA}) \to 0$. Then the $S(\theta)_{EA}$ becomes:

$$S(\theta)_{EA} = x^2 \frac{J_1(z)}{z},$$

which is nothing more than the scattering function in the Fraunhofer diffraction approximation.

(ii) Another domain where the integration can be performed for non-forward scattering is $x|m^2 - 1| \ll 1$. If this condition is satisfied one may approximate:

$$\exp[\omega^* \cos \gamma] \simeq 1 + \omega^* \cos \gamma.$$

The scattering function is then:

$$S(\theta)_{EA} = i(m^2 - 1)x \left[\cos z - \frac{\sin z}{z}\right] \bigg/ 2 \sin^2(\theta/2),$$

which, for $\theta = 0$, is nothing more than the scattering function for a homogeneous sphere in the Rayleigh–Gans–Debye approximation (RGDA). This approximation is dealt with in Chapter 4. For the moment, it suffices to note that the RGDA is known to be a good approximation in the domain:

$$|m - 1| \ll 1; \quad 2x|m - 1| \leq 1.$$

Since the scattering function in the EA reduces to the scattering function for the RGDA in its validity domain, it may be concluded that, despite the original premise $x \gg 1$ for the validity of the EA, in practice it can be a good approximation, at least for small-angle scattering, for arbitrary x as long as the condition $|m - 1| \ll 1$ is satisfied. This, indeed, has been verified by numerical comparisons for scattered intensities. The implication of this result is that cancellations must occur between errors arising from the scalar approximation and errors inherent in the EA. It may be pointed out here that for a brief period there was some confusion regarding the validity domain of the ADA when Liu et al. (1996) suggested that its accuracy was not sensitive to the condition $|m - 1| \ll 1$. However, ensuing research (Liu, 1998; Videen and Chýlek, 1998) clarified this intricacy, and it was established that the results are indeed sensitive to the condition $|m - 1| \ll 1$.

A difference in sign between the phase shift function in (3.30) and the corresponding van de Hulst (1957) expression may be noted. This difference is a result of the choice of sign in the representation of the incoming plane wave. In potential scattering – that is, in quantum mechanics problems – the usual choice for the incoming plane wave is $\exp[i(kz - \omega t)]$. In this book we have adopted this convention. The same convention has been adopted in books by Newton (1966), Bayvel and Jones (1981), Bohren and Huffman (1983), Mishchenko et al. (2002), and Babenko et al. (2003). The notation for a complex refractive index is then $m = n + in'$. On the other hand, van de Hulst (1957), Kerker (1969) and Kokhanovsky (2005) use the convention $\exp[-i(kz - \omega t)]$ for the incoming wave. The complex refractive index is then $m = n - in'$. The effect on the scattering function is that the scattering functions in this book are complex conjugates of the functions in van de Hulst (1957) or Kerker (1969). The scattered intensity and efficiency factors do not need to be changed. Physical processes are of course equivalently described in both representations. The implications of choosing a particular convention have been discussed in greater detail by Shifrin and Zolotov (1993).

The fact that $S(\theta)_{EA}$ is formally identical to $S(\theta)_{ADA}$ allows many a result obtained for the ADA to be used for the EA in a straightforward way. For forward scattering, integration in $S(0)$ can be evaluated in a closed form. Then, using the extinction theorem (3.7), one obtains:

$$Q_{EA}^{ext} = \left[2 - \frac{4\cos\beta_{EA}}{\rho_{EA}}\sin(\rho_{EA} - \beta_{EA})\exp(-\rho_{EA}\tan\beta_{EA}) + 4\left(\frac{\cos\beta_{EA}}{\rho_{EA}}\right)^2\cos(2\beta_{EA})\right.$$
$$\left. - 4\left(\frac{\cos\beta_{EA}}{\rho_{EA}}\right)^2\cos(\rho_{EA} - 2\beta_{EA})\exp(-\rho_{EA}\tan\beta_{EA})\right], \quad (3.42)$$

as the extinction efficiency. For a nonabsorbing sphere $\sin\beta_{EA} = 0$ and equation (3.42) reduces to:

$$Q_{EA}^{ext} = \left[2 - \frac{4}{\rho_{EA}}\sin\rho_{EA} + \frac{4}{\rho_{EA}^2}(1 - \cos\rho_{EA})\right]. \quad (3.43)$$

A typical variation of Q_{EA}^{ext} with x is shown in Figure 3.2. It can be seen that the extinction efficiency rises steeply as ρ_{EA} increases from 0 and reaches a maximum value at $\rho_{EA} = 4.1$. It then oscillates around a value of 2 with decaying amplitude. The period of oscillation is $\rho_{EA} = 2\pi$. For $|\rho_{EA}| \to \infty$, (3.43) gives $Q_{EA}^{ext} = 2$, which is the correct asymptotic value of the extinction. Note that the ripple structure disappears as either $m \to 1$ or as the imaginary part of the complex refractive index increases.

It has been noted that – although the EA as well as the ADA lead to the correct $x \to \infty$ limit of the extinction efficiency factor – the rate of approach to this limit is much faster than predicted by Mie theory. This difference can be attributed to the effect at the edge of the particle. Based on the work of Nussenzveig and Wiscombe (1980), this edge effect is included as an additional term in the extinction formula (Ackerman and Stephens, 1987):

$$Q_{AS}^{ext} = Q_{ADA}^{ext} + Q_{edge}, \quad (3.44)$$

where

$$Q_{edge} = 2x^{-2/3}. \quad (3.45)$$

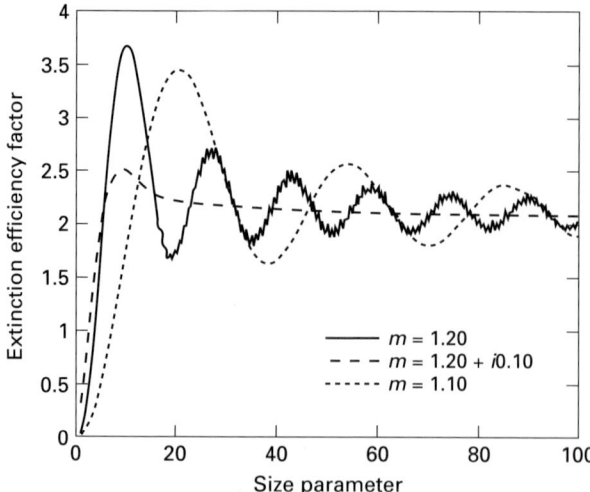

Figure 3.2. Typical variations in extinction efficiency curve with size parameter and refractive index.

The edge corrections are essentially wave optics corrections to geometrical optics results. A more general edge correction also includes a term describing the interference between surface waves that gives rise to the ripple structure in the extinction efficiency curve. But, this ripple structure is not of much interest in problems relating to scattering by a collection of particles because of averaging effects and also for particles where the condition $|n-1| \ll 1$ is well-satisfied or when the imaginary part of the refractive index is large. This can be seen in Figure 3.2.

Approaching the problem from a different perspective, an empirical recipe to improve the accuracy of Q_{ADA}^{ext} has been given by Klett (1984). His correction, for nonabsorbing spheres, can be expressed as:

$$Q_K^{ext} = \mathcal{D} Q_{ADA}^{ext}, \tag{3.46}$$

where

$$\mathcal{D} = 1.1 + \frac{(n-1.2)}{3}, \tag{3.47}$$

for n between 1 and 5 roughly. The above expression is just an *ad hoc* prescription and has no theoretical basis. For absorbing spheres, n in the above equation should be replaced by $|m|$. The $\mathcal{D}$ values obtained from (3.47) agree reasonably well with a similar m-dependent *ad hoc* prescription obtained later in (3.77).

An empirical formula for the extinction efficiency of a large dielectric sphere has been given by Walstra (1964). The result is:

$$Q_{WA}^{ext} = 2 - \frac{16m^2 \sin\rho}{(m+1)^2 \rho} + 4\frac{1 - m\cos\rho}{\rho^2} + 7.53\frac{\bar{z}-m}{\bar{z}+m}x^{-0.772}, \tag{3.48}$$

where

$$\bar{z} = \left[(m^2-1)(6x/\pi)^{(2/3)} + 1\right]^{1/2}.$$

The equation (3.48) gives values correct to within 1% for $\rho > 2.4$ and $1 < m \leq 1.25$. Even for higher values of m the formula is useful, provided:

$$x > \frac{\pi}{4(\cot^{-1} m)^3}.$$

If $m \to 1$, $Q_{WA}^{ext} \to Q_{ADA}^{ext}$ and if $x \to \infty$, $Q_{WA}^{ext} \to 2$, as expected. A semi-empirical formula for $S(0)$ was also derived by Walstra (1964) on the same lines. The result is:

$$S(0)_{WA} = \frac{1}{2}x^2 - i\frac{2m^2 x \exp(-i\rho)}{(m+1)^2(m-1)} + \frac{1 - m\exp(-i\rho)}{4(m-1)^2}$$

$$+ (1.88 - 1.05i)\frac{\bar{z} - m}{\bar{z} + m} x^{1.228}. \quad (3.49)$$

This result is expected to give values correct to within 1% for $\rho > 3$, and $1 < m \leq 1.25$.

Some other alternative forms for (3.43) that have proved to be useful are (see, e.g., Fymat, 1978):

$$Q_{EA}^{ext} = 2 + 4\frac{d}{d\rho}\left(\frac{\cos\rho - 1}{\rho}\right),$$

$$Q_{EA}^{ext} = 2\left(1 + \frac{2}{\rho_{EA}^2}\right) + 2\left(\frac{2\pi}{\rho_{EA}}\right)^{1/2} J_{-3/2}(\rho_{EA}),$$

$$Q_{EA}^{ext} = 4\left(\frac{2\pi}{2\rho_{EA}^{ext}}\right)^{1/2} H_{3/2}(\rho_{EA}),$$

$$Q_{EA}^{ext} = \tfrac{1}{2}\rho_{EA}^2 F(1; 3, 3/2, -\tfrac{1}{4}\rho_{EA}^2),$$

where J is the spherical Bessel function of the first kind, and H and F are Struve and hypergeometric functions. These forms are particularly useful while performing analytic inversion to obtain the size distribution of scatterers from the extinction spectrum of a collection of particles.

Let us now scrutinize $S(\theta)$ for non-forward scattering. It is clear that, in general, it is not possible to evaluate $S(\theta)_{EA}$ analytically in a closed form. However, it is found that it is possible to express the scattering function in terms of known functions by means of a series expansion for real values of refractive index ($n' = 0$) (van de Hulst, 1957). Two separate series expansions are arrived at in two separate domains. For small values of ρ_{EA} an expansion in powers of ρ_{EA} yields:

$$\text{Re } S(\theta)_{EA} = x^2\left[\left(\frac{\rho_{EA}}{z}\right)^2 J_2(z) - \frac{\rho_{EA}^4}{1.3}\frac{1}{z^3}J_3(z) + \frac{\rho_{EA}^6}{1.3.5}\frac{1}{z^5}J_4(z) + \cdots\right].$$

For $\rho_{EA} > z$, one obtains:

$$\text{Re } S(\theta)_{EA} = x^2\left[\frac{1}{z}J_1(z) + \frac{\rho_{EA}}{y^{3/2}}\sqrt{\pi/2}N_{3/2}(y) + \frac{1}{\rho_{EA}^2}J_0(z) + \frac{1.3}{\rho_{EA}^4}zJ_1(z) + \frac{1.3.5}{\rho_{EA}^6}z^2 J_2(z) + \cdots\right].$$

This series expansion for Re $S(\theta)_{EA}$ is obtained by making a doubly infinite expansion in powers of ρ_{EA} and z.

As for the imaginary part of the $S(\theta)_{EA}$, integrals can be evaluated analytically to yield:

$$\text{Im } S(\theta)_{EA} = -\rho_{EA} \frac{x^2}{y^2} \sqrt{\frac{\pi y}{2}} J_{3/2}(y),$$

where

$$y^2 = \rho_{EA}^2 + z^2.$$

For the real part of the scattering function, two separate series expansions are arrived at in two separate domains.

The absorption efficiency in the EA may be obtained following the derivation of absorption efficiency in the ADA by van de Hulst (1957). This gives:

$$Q_{EA}^{abs} = \frac{1}{\pi a^2} \int d\mathbf{b}[1 - \exp(-\text{Im}\chi_{EA})]. \tag{3.50}$$

For a homogeneous sphere, integrations in (3.50) can be performed in a straightforward manner leading to:

$$Q_{EA}^{abs} = 2K(-2\kappa_{EA}) = 1 + \frac{e^{-2\kappa_{EA}}}{\kappa_{EA}} + \frac{(e^{-2\kappa_{EA}} - 1)}{2\kappa_{EA}^2}. \tag{3.51}$$

When $\kappa_{EA} \gg 1$, the absorption efficiency is close to 1. That is, all the rays incident on the sphere are absorbed. In addition, an equal amount of light is diffracted (scattered) resulting in the so-called "extinction paradox". It should be mentioned here that the exact calculations of the absorption efficiency for nonspherical particles show that its value can be larger than unity because of the tunneling effect which allows incident radiation outside the projected area to be absorbed.

The formulation for the extinction efficiency factor further simplifies to:

$$Q_{EA}^{ext} = 2 + \frac{4e^{-\kappa_{EA}}}{\kappa_{EA}}(1 + \kappa_{EA}) - \frac{4}{\kappa_{EA}^2},$$

if the particle is such that $\rho_{EA} \ll 1$ while $\kappa_{EA} \gg 1$.

It can be shown easily that the expressions for the extinction and absorption efficiency factors for $|\rho_{EA}^*| \ll 1$ become:

$$Q_{EA}^{ext} = \tfrac{4}{3}\kappa_{EA} + \tfrac{1}{2}(\rho_{EA}^2 - \kappa_{EA}^2),$$

and

$$Q_{EA}^{abs} = \tfrac{4}{3}\kappa_{EA} - \kappa_{EA}^2.$$

The scattering efficiency is then:

$$Q_{EA}^{sca} = \tfrac{1}{2}(\rho_{EA}^2 + \kappa_{EA}^2) = \tfrac{1}{2}|\rho_{EA}^*|^2.$$

Clearly, extinction and absorption are first-order effects, while scattering is a second-

3.4.2 Derivation from the Mie solutions

The Mie solutions are exact solutions that describe the scattering of light by a homogeneous sphere of arbitrary size and refractive index. One gets two scattering functions, $S_1(\theta)$ and $S_2(\theta)$, corresponding to perpendicular and parallel polarizations, respectively. They are given by the following expressions:

$$S_1(\theta) = \sum_{l=1}^{\infty} \frac{2l+1}{l(l+1)} [a_l \pi_l(\cos\theta) + b_l \tau_l(\cos\theta)], \quad (3.52)$$

and

$$S_2(\theta) = \sum_{l=1}^{\infty} \frac{2l+1}{l(l+1)} [a_l \tau_l(\cos\theta) + b_l \pi_l(\cos\theta)], \quad (3.53)$$

where

$$\pi_l(\cos\theta) = \frac{dP_l(\cos\theta)}{d(\cos\theta)},$$

and

$$\tau_l(\cos\theta) = \cos\theta \, \pi_l(\cos\theta) - \sin\theta \left[\frac{d\pi_l(\cos\theta)}{d(\cos\theta)}\right],$$

with $P_l(\cos\theta)$ as the Legendre polynomial. The scattering coefficients a_l and b_l can be expressed as:

$$a_l = \frac{1 - e^{2i\alpha_l}}{2} = \frac{mu_l(mx)u_l'(x) - u_l(x)u_l'(mx)}{mu_l(mx)\zeta_l'(x) - u_l'(mx)\zeta_l(x)}, \quad (3.54)$$

$$b_l = \frac{1 - e^{2i\beta_l}}{2} = \frac{u_l(mx)u_l'(x) - mu_l(x)u_l'(mx)}{u_l(mx)\zeta_l'(x) - mu_l'(mx)\zeta_l(x)}, \quad (3.55)$$

and

$$\tan\alpha_l = \frac{u_l'(mx)u_l(x) - mu_l(mx)u_l'(x)}{u_l'(mx)v_l(x) - mu_l(mx)v_l'(x)}, \quad (3.56)$$

$$\tan\beta_l = \frac{mu_l'(mx)u_l(x) - u_l(mx)u_l'(x)}{mu_l'(mx)v_l(x) - u_l(mx)v_l'(x)}, \quad (3.57)$$

with

$$u_l(x) = zj_l(x), \quad v_l(x) = zv_l(x) \quad \text{and} \quad \zeta_l(x) = zh_l^{(1)}(x),$$

as the Riccati–Bessel, Riccati–Neumann and Riccati–Hankel functions and $h_l^{(1)}(x) = j_l(x) - n_l(x)$. Similar relations also apply when the argument x changes to mx.

It can easily be shown that in the limit:

$$x \to \infty, \quad |mx| \to \infty \quad \text{and} \quad \frac{n'^2 x}{n} \ll 1,$$

the coefficients α_l and β_l may be approximated as (Bourrley et al., 1991):

$$e^{2i\alpha_l} = e^{-2i(xf - x'f')} \frac{1 + r_\perp e^{2n'x} e^{-2ix'f'}}{1 - r_\perp e^{-2n'x} e^{2ix'f'}} e^{-2n'x}, \qquad (3.58)$$

$$e^{2i\beta_l} = e^{-2i(xf - x'f')} \frac{1 + r_\| e^{2n'x} e^{-2ix'f'}}{1 - r_\| e^{-2n'x} e^{2ix'f'}} e^{-2n'x}, \qquad (3.59)$$

where

$$f = \cos\gamma - \left[\gamma - \frac{\pi}{2}\right]\sin\gamma,$$

$$f' = \cos\gamma' - \left[\gamma' - \frac{\pi}{2}\right]\sin\gamma',$$

and $x' = nx$. The Fresnel reflection coefficients $r_\|$ and $r_\perp$ are given by:

$$r_\|(\gamma) = \frac{m\cos\gamma - \cos\gamma'}{m\cos\gamma + \cos\gamma'}, \qquad (3.60)$$

and

$$r_\perp(\gamma) = \frac{\cos\gamma - m\cos\gamma'}{\cos\gamma + m\cos\gamma'}. \qquad (3.61)$$

The angle of incidence γ (Figure 3.1) is related to γ' through the relation:

$$2x\sin\gamma = 2x'\sin\gamma' = (2l + 1). \qquad (3.62)$$

The expressions (3.58) and (3.59) are generalizations of the corresponding expressions for a nonabsorbing homogeneous sphere ($n' = 0$) derived in most books dealing with light scattering (see, e.g., Newton, 1966; or van de Hulst, 1957). The added feature in the derivation of Bourrley et al. (1991) consists in incorporating complex argument Hankel functions and writing it as:

$$H^{(1)}_{l+\frac{1}{2}}(mx) \simeq e^{-n'x} H^{(1)}_{l+\frac{1}{2}}(nx)$$

for a weakly absorbing sphere ($n'^2 x/n \ll 1$). Thus, $(xf - x'f')$ can be expressed as:

$$xf - x'f' = -x\eta + x(\gamma' - \gamma)\sin\gamma, \qquad (3.63)$$

where

$$\eta = n\cos\gamma' - \cos\gamma, \qquad (3.64)$$

which may also be rewritten as:

$$\eta = [(1 + \xi)^{1/2} - 1]\cos\gamma, \qquad (3.65)$$

with

$$\xi = \frac{(n^2 - 1)}{\cos^2\gamma}. \qquad (3.66)$$

Since $|n-1| \ll 1$, ξ is also much less than unity and one can expand $(1+\xi)^{1/2}$ in powers of ξ. The leading term in the limit $\xi \to 0$ or $n \to 1$ approximates η as:

$$\eta = \frac{(n^2 - 1)}{2\cos\gamma}.$$

Similarly, $(\gamma' - \gamma)$ may be approximated as:

$$\gamma' - \gamma = (n^2 - 1)\tan\gamma/2.$$

The equation (3.63) then gives:

$$xf - x'f' = -x(n^2 - 1)\cos\gamma/2,$$

as the leading $(n^2 - 1)$ term. For large x and small scattering angles ($\theta \ll 1/x$) one can introduce the approximations:

$$\pi_l(\cos\theta) \simeq \tau_l(\cos\theta) \simeq \frac{l(l+1)}{2} J_0\left[(l+\tfrac{1}{2})\theta\right],$$

and

$$\sum_{l=1}^{l=\infty} \simeq x \int_0^{\pi/2} \cos\gamma \, d\gamma.$$

In addition, if $n \to 1$, $r_\parallel$ and $r_\perp$ go to 0. It is then straightforward to see that (3.52) as well as (3.53) reduce to $S(\theta)_{EA}$. It may be mentioned at this point that the assumption:

$$\frac{|n^2 - 1|}{\cos^2\gamma} \ll 1 \qquad (3.67)$$

used above is not valid near $\gamma = \pi/2$ (i.e., near grazing incidence). Nevertheless, the approximation is a good one because the contribution of such rays is small (van de Hulst, 1957).

3.4.3 Relationship with the anomalous diffraction approximation

A systematic expansion of $(xf - x'f')$ in powers of $(n^2 - 1)$ up to order $(n^2 - 1)^3$ gives (Sharma, 1992):

$$-2i(xf - x'f') = i(n^2 - 1)x\cos\gamma\Big[1 - \tfrac{1}{4}(n^2 - 1)(1 - \tan^2\gamma)$$
$$+ \tfrac{1}{8}(n^2 - 1)^2(1 - 2\tan^2\gamma) - \tfrac{1}{24}(n^2 - 1)^2 \tan^4\gamma\Big]. \qquad (3.68)$$

The first term on the right-hand side of (3.68) is the phase obtained in the EA. Clearly, the EA phase should be viewed as an $n \to 1$ approximation for fixed $(n^2 - 1)x$. Corrections to the EA phase in (3.68) may now be contrasted with the phase obtained by translating the Wallace results of potential scattering (Section 2.8.1) to optical scattering. The Wallace phase, to order $(n^2 - 1)^3$, then reads:

$$i(n^2 - 1)x\cos\gamma - ix\frac{(n^2 - 1)^2}{4}(1 - \tan^2\gamma)\cos\gamma + ix\frac{(n^2 - 1)^3}{8}(1 - 2\tan^2\gamma)\cos\gamma. \tag{3.69}$$

It is gratifying to note that the first-order correction in (3.68) is identical to the first-order correction of Wallace. The second-order correction, however, agrees only partially. It may be mentioned here that the expression (3.69) does not include contributions of w_2-type. That is, w_2-type corrections seem to be absent in optical scattering. However, there is a possibility that these may be related to $r_\parallel$ and $r_\perp$ terms in (3.58) and (3.59).

Regrouping the terms in (3.68) we can rewrite it as:

$$-2i(xf - x'f') = 2i\left[\tfrac{1}{2}(n^2 - 1) - \tfrac{1}{8}(n^2 - 1)^2 + \tfrac{1}{16}(n^2 - 1)^3\right]x\cos\gamma$$

$$-\frac{i}{4}x(n^2 - 1)^2\tan\gamma\sin\gamma$$

$$-\frac{i}{4}x(n^2 - 1)^3\tan\gamma - \frac{i}{24}x(n^2 - 1)^3\sin\gamma\tan^3\gamma. \tag{3.70}$$

If the term in square brackets is approximated as:

$$\left[[1 + (n^2 - 1)]^{1/2} - 1\right] \simeq (n - 1),$$

then equation (3.70) becomes:

$$-2i(xf - x'f') = 2ix(n - 1)\cos\gamma + \frac{i}{4}x(n^2 - 1)^2\tan\gamma\sin\gamma$$

$$-\frac{i}{4}x(n^2 - 1)^3\tan\gamma\sin\gamma - \frac{i}{24}x(n^2 - 1)^3\tan^3\gamma\sin\gamma. \tag{3.71}$$

The first term on the right-hand side of this equation is the ADA phase. Note that the corrections tend to 0 as either $|n - 1| \to 0$ or as $\gamma \to 0$. This equation also tells us that the ADA phase is a good approximation to the exact phase if:

$$\frac{\sin^2\gamma}{\cos\gamma} \ll \frac{1}{\rho|n - 1|}.$$

This means that as $n \to 1$ for fixed ρ, the domain of γ values over which this approximation is valid increases and, hence, the ADA improves. Thus, while it is true that the validity of the ADA phase improves as $n \to 1$, nevertheless it is also true that its validity is not limited to $|n - 1| \ll 1$. For $\gamma = 0$ the approximation of phase is valid for arbitrary n.

It is evident from the above discussion that the domain of γ values where the phase approximation is good is quite small if the condition $|n - 1| \ll 1$ is not satisfied.

But, the very fact that the phase approximation is good near $\gamma = 0$ is significant. This is because the main contribution of the refraction term to near forward scattering comes from the region near $\gamma = 0$ (see, e.g., van de Hulst, 1957). Contributions to $\theta = 0$ from noncentral and nongrazing rays are of little importance and may, therefore, be ignored. Since $-r_\perp = r_\parallel = (m-1)/(m+1)$ for near central incidence, it may be concluded that the ADA should be a good approximation if $|n-1|^2 \ll |n+1|^2$. This is a crucial observation and explains why the anomalous diffraction approximation is a reasonably good approximation even for n as large as 2.

Finally, we also note that, though the ADA phase contains parts of the corrections to the EA phase, it need not necessarily be a better approximation because the errors (terms neglected) in both the approximate phases are of the same order. However, for large ρ, where the important contributions of the refraction term to forward scattering comes from γ values near 0, corrections to the ADA phase are much smaller and, hence, the ADA may be expected to give better results. Indeed, in a numerical comparison of the ADA and the EA for forward scattering from a homogeneous sphere it was noted (Debi and Sharma, 1979) that the ADA gives superior results for $\rho > 4$ (see Table 3.3).

Derivation of the EA from the Mie theory and discussion of its relationship with the ADA is limited in this section by the condition $n'^2 x/n \ll 1$ and, hence, essentially to weakly absorbing spherical particles. But, this restriction is a shortcoming of the derivation and is not an intrinsic limitation. The validity of equation (3.69), which is obtained by translating potential scattering results to optical scattering, is not restricted to a weakly absorbing sphere. The conclusions in (3.4.2) and (3.4.3) are thus valid for arbitrary absorption.

Table 3.3. Percent error in various approximate methods in $i(0)$ for a homogeneous sphere of refractive index 1.05.

x	ρ_{EA}	$\rho_{EA}(m^2-1)/4$	EA	FCI	ADA	FCII	MFCI
1.0	0.1025	2.63×10^{-3}	−3.07	−8.43	1.88	−7.59	0.13
3.0	0.3075	7.88×10^{-3}	1.37	−3.79	6.09	−2.97	−0.69
5.0	i0.5125	1.31×10^{-2}	2.58	−2.56	7.20	−1.76	−0.68
10.0	1.025	2.63×10^{-2}	3.67	−1.65	8.04	−0.84	−0.70
20.0	2.05	5.25×10^{-2}	4.90	−1.37	8.35	−0.50	−0.89
30.0	3.075	7.88×10^{-2}	6.60	−1.33	8.45	−0.39	−1.01
40.0	4.10	1.05×10^{-1}	9.11	−1.31	8.55	−0.28	−1.07
50.0	5.125	1.31×10^{-1}	12.65	−1.19	8.65	−0.17	−1.0
60.0	6.15	1.58×10^{-1}	16.65	−0.89	8.67	−0.14	−0.73
70.0	7.175	1.84×10^{-1}	15.17	−0.67	8.10	−0.77	−0.54
80.0	8.20	2.10×10^{-1}	8.20	−0.67	6.80	−2.20	−1.94
90.0	9.225	2.35×10^{-1}	−2.60	−3.44	6.65	−2.36	
100.0	10.25	2.63×10^{-1}	2.98	−3.73	7.04	−1.93	

3.4.4 Corrections to the eikonal approximation

When converted to optical scattering, the first-order, corrected scattering function of the Wallace can be written (Sharma *et al.*, 1982) as:

$$S(\theta) = x^2 \int_0^{\pi/2} J_0(z \sin \gamma)[1 - e^{i\chi_{EA} + i\tau_1}] \cos \gamma \sin \gamma \, d\gamma, \tag{3.72}$$

where

$$\tau_1 = -\frac{x(m^2 - 1)^2 \cos \gamma}{4}(1 - \tan^2 \gamma).$$

A difficulty arises while evaluating (3.72). At $\gamma = \pi/2$, correction to the eikonal phase diverges. This is essentially a consequence of the sharp cut-off at the boundary of the scatterer. The eikonal expansion of Wallace, derived for infinitely often differentiable potentials, thus need not hold even at the zeroth-order level for optical scattering. But, because numerical comparisons confirm that the EA is a good approximation, it is customary to ignore the mathematical problems associated with discontinuous behavior at the boundaries. Thus, one may either evaluate the integration ignoring the contributions of rays near $\gamma = \pi/2$ or may consider only that part of the correction term which is free of the divergence problem. An approximate form which is free of this problem in the domain:

$$\frac{x|m^2 - 1|^2}{4} \ll 1, \tag{3.73}$$

is:

$$S(\theta)_{FCI} = x^2 \int J_0(z \sin \gamma)[1 - e^{i\chi_{EA}}(1 + i\tau_1)] \sin \gamma \cos \gamma \, d\gamma. \tag{3.74}$$

Recall that $\chi_{EA} = x(m^2 - 1) \cos \gamma$. The abbreviation FC in the subscript refers to the first-order correction. Another similar correction can be based on (3.71). The leading phase term here is the ADA phase. The result is:

$$S(\theta)_{FCII} = x^2 \int_0^{\pi/2} J_0(z \sin \gamma)\left[1 - e^{i\chi_{ADA}}\left(1 + \frac{ix(m^2 - 1)}{4 \cos \gamma} \sin^2 \gamma\right)\right] \times \sin \gamma \cos \gamma \, d\gamma, \tag{3.75}$$

where $\chi_{ADA} = 2(m - 1) \cos \gamma$. For forward scattering, these can be evaluated analytically, to yield:

$$S(0)_{FCI} = m^2 S(0)_{EA} + \frac{x^2}{4}(m^2 - 1)[e^{-ix(m^2 - 1)} - 1], \tag{3.76}$$

and

$$S(0)_{FCII} = C(m)S(0)_{ADA}, \tag{3.77}$$

with

$$C(m) = 1 + 0.25(m + 1)^2(m - 1)(2 - m^2). \tag{3.78}$$

The correction given by (3.77) clearly preserves the simplicity of the unmodified approximation.

A generalized form of the eikonal approximation (GEA) has been examined by Chen (1989) for the problem of scattering of light by a homogeneous sphere. The scattering function in the GEA can be obtained simply by replacing $U(r)$ in (2.60) as $U(r) = k^2(1 - m^2)\Theta(a - r)$. The scattering function in the GEA then takes the following form:

$$S(\theta)_{GEA} = \frac{ik}{4\pi}(1 - \bar{\delta})S_{B1} - (m^2 - 1)\alpha_0\bar{\delta}^2 \, k^2 I, \tag{3.79}$$

where the first Born term S_{B1} is given by:

$$S_{B1}(\theta) = -4x^2(m^2 - 1)a \int_0^{\pi/2} \sin^2 \gamma \cos \gamma J_0(z \sin \gamma) \, d\gamma$$

and

$$I = a^2 \int_0^{\pi/2} J_0(z \sin \gamma)\left[e^{i\rho^*_{GEA} \cos \gamma} - 1\right] \cos \gamma \sin \gamma \, d\gamma,$$

with

$$\rho^*_{GEA} = x(m^2 - 1)/\alpha_0\bar{\delta}. \tag{3.80}$$

In the above equations, $\alpha_0 = \alpha/k$ is a dimensionless quantity and $\bar{\delta}$ is a function of x and $m - 1$. In potential scattering, the parameters α and δ were determined by requiring that the scattering amplitude reproduces the real and imaginary part of the second Born term correctly (Section 2.8.3). In scattering by optically soft particles, parameter determination is based on the following two criteria:

(i) The change in phase of incident radiation in propagation through the medium is $2k(m - 1)\sqrt{a^2 - b^2}$.

(ii) The edge effects due to the sharp boundary are recovered.

The first condition gives the following relation:

$$\rho^*_{GEA} = x(m^2 - 1)/\alpha_0\bar{\delta} = \rho^*_{ADA}, \tag{3.81}$$

that is:

$$\bar{\delta} = (m + 1)/2\alpha_0. \tag{3.82}$$

The second condition gives:

$$\alpha_0 = \frac{m + 1}{2} - \frac{3i}{8}\left[\frac{1}{x} - \frac{2}{\rho^*_{GEA}}\left(\frac{a_1}{x^{2/3}} - \frac{a_2}{x^{4/3}}\right)\right], \tag{3.83}$$

where $a_1 = 2 + 2.4i, a_2 = 2 + 6i$. For forward scattering, the integrals in (3.79) can easily be evaluated, leading to an analytic expression for the scattering function and, hence, for the extinction efficiency.

In the calculation of phase shift the GEA assumes that light passes undeviated through the medium. The modified GEA (MGEA) corrects it by arguing that – although light travels in a straight line inside the medium – the direction of travel is no longer along the incident direction. Instead, it is along the direction of the average of both incident and scattered directions. That is, it is assumed that the light scattered at an angle θ is scattered twice. At each boundary it suffers a deviation $\theta/2$. Propagation inside the medium is along a straight line. Thus, the phase shift with and without the medium is $2k(m - \cos\theta/2)\sqrt{(a^2 - b^2)}$. Thus, in the MGEA (Chen and Smith, 1992; Chen, 1993):

$$\rho^*_{MGEA} = \frac{x[m^2 - 1]}{\alpha_0 \bar{\delta}} = 2x[m - \cos(\theta/2)],$$

or

$$\bar{\delta} = \frac{(m^2 - 1)}{2\alpha_0[m - \cos(\theta/2)]},$$

is the modified relation between α_0 and $\bar{\delta}$.

It can easily be verified that for forward scattering the integrations in the expression for the scattering function can easily be evaluated, and one obtains an analytic expression for the forward scattering function. Further, it can also be seen that the generalized eikonal results reduce to the standard EA for $\bar{\delta} = \alpha_0 = 1$.

Integrals in the scattering function can be evaluated for all angles for a large sphere – that is, when $\rho^*_{MGEA} \gg 1$. This yields:

$$S(\theta)^{LS}_{MGEA} = (m^2 - 1)x^3 \left[-i(1-\delta)\frac{j_1(z)}{z} + \frac{\delta J_1(z)}{z\rho^*_{MGEA}} + \frac{\delta}{y_1^2}\left(ie^{iy_1} + \frac{1 - e^{iy_1}}{y_1}\right) \right], \quad (3.84)$$

where

$$y_1 = [z^2 + \rho^{*2}_{MGEA}]^{1/2},$$

and α_0 is redefined as:

$$\alpha_0 = \frac{m+1}{2} - \frac{3i}{8x},$$

by ignoring the higher order terms in (3.83). The superscript LS refers to large scatterers. Equation (3.84) further simplifies to:

$$S(\theta)^{LS}_{MGEA} \approx (m^2 - 1)x^3 \delta \left[\frac{J_1(z)}{\rho^*_{MGEA} z} - \frac{3j_1(z)}{4x(m+1)z} + i\frac{e^{iy_1}}{y_1^2} + \frac{1 - e^{iy_1}}{y_1^3} \right], \quad (3.85)$$

for small θ and to lowest order in $1/x$. This simplified expression for scattered intensity leads to simple relations between the positions of minima in the scattering pattern and size of the scatterer and are given in Chapter 5.

From the point of view of widening the angular domain of the validity of the EA to larger angles, Perrin and Chiappetta (1985) have proposed a modification of the EA which is referred to as the "eikonal picture" (EP). The explicit small-angle approximation made for arriving at the two-dimensional scattering amplitude is not employed here. The EP is simply a translation of the result (2.25) of potential scattering to optical scattering. Thus, the scattering function in the EP is written as:

$$S(\theta)_{EP} = -\frac{ik^3}{2} \int d\mathbf{b}\, dz\, e^{i\mathbf{q}\cdot\mathbf{b}}[m^2(\mathbf{b}, z) - 1]$$

$$\times \exp\left[2ikz \sin^2(\theta/2) + \frac{ik}{2}\int_{-\infty}^{z(\mathbf{b})} [m^2(\mathbf{b}, z') - 1]\, dz' \right]. \quad (3.86)$$

For $\theta = 0$, $\exp(2ikz \sin\theta/2) = 1$, the EP is then identical with the EA. For a homogeneous sphere, z integration in $S(\theta)_{EP}$ can be performed analytically. This yields:

$$S(\theta)_{EP} = -k^3(m^2 - 1) \int_0^a b\, db\, J_0(kb \sin\theta) \left[\frac{e^{i[q_z + k(m^2-1)]\sqrt{a^2-b^2}} - e^{-iq_z\sqrt{a^2-b^2}}}{2q_z + k(m^2-1)} \right]. \quad (3.87)$$

At first sight it might appear from (3.87) that when $\text{Im}|m^2 - 1|$ is negligible and $4\sin^2(\theta/2) + \text{Re}|m^2 - 1| = 0$, the integrand diverges. But, a closer look at the integrand shows that the apparent divergences are in fact spurious. A variant similar to this was suggested by Berlad (1971) in the context of scattering of spin-1/2 particles.

An alternative form in which (3.87) is sometimes expressed is:

$$S(\theta)_{EP} = \frac{ix^2(m^2 - 1)}{U} \int_0^1 y\, dy\, J_0(yx \sin\theta) \sin\left(Ux\sqrt{1-y^2}\right) e^{iUx\sqrt{1-y^2}} \quad (3.88)$$

where

$$U = \frac{[2q_z + k(m^2 - 1)]}{2k}, \quad (3.89)$$

and, as before, $q_z = 2k \sin^2(\theta/2)$. Equation (3.88) has been obtained from (3.87) by making the change of variable $b = ay$. This equation may be contrasted with the scattering function in the Wentzel–Kramers–Brillouin approximation (WKBA) (Klett and Sutherland, 1992; Shepelevich et al., 1999):

$$S(\theta)_{WKB} = \frac{ix^2(m^2 - 1)}{m - \cos\theta} \int_0^1 y\, dy\, J_0(yx \sin\theta) \sin\left[(m - \cos\theta)x\sqrt{1-y^2}\right] e^{i(m-\cos\theta)x\sqrt{1-y^2}}. \quad (3.90)$$

Note that if $(m^2 - 1)$ is replaced by $2(m - 1)$ in (3.89), U is nothing more than $(m - \cos\theta)$. Equation (3.90) is then the same as (3.88).

When the refractive index satisfies the condition $\text{Im}(mx) \geq 1$, it is easy to see that the first term in the square brackets in (3.87) may be neglected. Then, the scattering function can be written as:

$$S(\theta)_{MEP} = k^2 \hat{r}(\theta) \int_0^a b\, db\, J_0(kb \sin\theta) e^{2ik \sin^2(\theta/2)\sqrt{a^2-b^2}}, \quad (3.91)$$

where

$$\hat{r}(\theta) = \frac{k(m^2 - 1)}{2q_z + k(m^2 - 1)}.$$

The subscript *MEP* indicates modified EP. For near forward scattering the assumption of a highly absorbing sphere means that the diffraction and the first reflective part together give a reasonable description of the scattering phenomenon. For near forward scattering (3.91) leads to diffractive scattering. For near backward scattering (3.91) gives the correct geometrical optics result if (Bourrley et al., 1996):

$$\hat{r}(\theta) = r_\perp(\theta),$$

where $r_\perp(\theta)$ is the Fresnel reflection coefficient for the perpendicular component of the electric field and is given by (3.61). With this replacement – that is, the replacement of

$\hat{r}(\theta)$ in (3.91) by $r_\perp(\theta)$ – the formula reproduces two main components of the scattering pattern and is valid in the forward and the backward scattering directions. The integral in (3.91) can be evaluated analytically in a series form in the same way as for the $S(\theta)_{EA}$.

A two-wave WKBA has been developed by Klett and Sutherland (1992). It is obtained by approximating $\psi(\mathbf{r})$, the field inside the scattering region, as:

$$\psi(\mathbf{r}) = e^{ik(m-1)\sqrt{a^2-b^2}} \left(e^{ikmz} + \frac{m-1}{m+1} e^{ikm(2\sqrt{a^2-b^2}-z)} \right). \tag{3.92}$$

The two-wave WKBA differs from the usual WKBA in that it allows for reflection from the back face of the particle. The first term on the right-hand side of (3.92) is the usual WKBA. The second term is the result of reflection from the back face of the particle. For unpolarized light the phase function in this approximation can be cast in the following form:

$$p(\theta) = \frac{2(1 + \cos^2\theta)|H_1 + \exp(i\rho_1)RH_2|^2}{\int_0^\pi (1 + \cos^2\theta)|H_1 + \exp(i\rho_1)RH_2|^2 \sin\theta\, d\theta}, \tag{3.93}$$

where

$$H_1(\theta) = \int_0^1 y\, dy\, J_0(yx\sin\theta) \sin\left[x(m - \cos\theta)\sqrt{1-y^2}\right] \exp\left[ix(m-1)\sqrt{1-y^2}\right],$$

$$H_2(\theta) = H_1(\pi - \theta),$$

and

$$R = \frac{1-m}{m+1},$$

with $\rho_1 = 2mx$. The single-wave WKBA is obtained by setting $H_2 = 0$ in equation (3.93) or, equivalently, in the second term on the right-hand side of (3.92).

The approximate expressions for the extinction and absorption efficiency factors in the ADA can be derived from the rigorous expressions of electrodynamics. The general expressions for extinction and absorption efficiency factors from electrodynamics can be written (see, e.g., Bohren and Huffman, 1983; Mishchenko et al., 2002; Yang et al., 2004), and the ADA can be incorporated in these relations by introducing it at the level of an electric field. This yields (Yang et al., 2004):

$$Q_{ADA}^{ext} = \text{Re}\left[\frac{m+1}{P} \int\int_P [1 - e^{ikl(m-1)}]\, dP\right], \tag{3.94}$$

and

$$Q_{ADA}^{abs} = \frac{n}{P} \int\int_P [1 - e^{-kln'}]\, dP, \tag{3.95}$$

where P is the projected area of the scattering object on a plane perpendicular to the incident direction and l is the geometric length traveled by the ray inside the scatterer. Equations (3.94) and (3.95) differ from corresponding, standard ADA efficiency factor expressions by multiplicative factors $(m + 1)/2$ and n, respectively. The asymptotic extinction efficiency factor, therefore, approaches a value $(m + 1)$ when the

Sec. 3.4] Scattering by a homogeneous sphere 41

particle is large and strongly absorptive. This does not agree with the correct value of the extinction efficiency factor, which is 2. To ensure that (3.94) and (3.95) lead to their correct asymptotic values, these equations have been modified empirically, resulting in the following expressions for the extinction and absorption efficiency factors:

$$Q_{ADA}^{ext} = \text{Re}\left\{\frac{2}{\bar{P}}\iint_P \left[1 - e^{-ikl(m-1)}\right] dP + e^{-\epsilon_1 V/\bar{P}}(m-1)\iint_P \left[1 - e^{-ikl(m-1)}\right] dP\right\}, \tag{3.96}$$

and

$$Q_{ADA}^{abs} = \frac{1}{\bar{P}}\iint_P \left[1 - e^{-kln'}\right] dP + \frac{1}{\bar{P}}e^{-\epsilon_2 V/\bar{P}}\iint_P \left[1 - e^{-kln'}\right] dP, \tag{3.97}$$

where V and $\bar{P}$, respectively, are the particle volume and orientation-averaged projected area. The constants ϵ_1 and ϵ_2 are tuning factors determined from comparison with corresponding exact solutions. The ϵ_1 and ϵ_2 determined for spherical particles have been used for nonspherical particles too and have been found to work well. The errors in the efficiency factors obtained from these expressions are much less than the errors in their conventional counterparts.

3.4.5 Numerical comparisons

To delineate the validity domains of the EA and the ADA, numerical comparisons for the forward scattered intensity $i(0)$ have been performed by Debi and Sharma (1979) for a nonabsorbing dielectric sphere. Since both these approximations are known to be small-angle approximations, the percent errors in forward scattered intensity were examined. A typical set of results is shown in Table 3.3. The percent error has been defined as:

$$\text{Percent error} = \frac{[i(0)_{exact} - i(0)_{approximate}] \times 100}{i(0)_{exact}}. \tag{3.98}$$

It is clear from the comparison in the table that the EA is superior to the ADA in the domain $x \geq 1.0$ and $\rho_{EA} \leq 4.0$. This observation is important because it relates to intermediate size particles. For higher values of ρ_{EA} the ADA gives more consistent results. On the other hand, for small-size particles the Rayleigh or the Rayleigh–Gans approximations are more useful. It is interesting to note that the value of $\rho_{EA} \simeq 4$ corresponds to the first maximum in the extinction curve.

As expected, the EA improves as $n \to 1$. But, unexpectedly, one finds that as x increases for a fixed n, the EA results do not improve continuously. The errors oscillate around the correct value. The reasons suggested to explain this behavior are as follows:

(*i*) Because of the energy dependence of the equivalent potential, the condition $U_0/k^2 \ll 1$ becomes $|n^2 - 1| \ll 1$, which is independent of the incident wave number. The approximation, therefore, improves as $n \to 1$, but does not show consistent improvement as k and, hence, x increases.

Table 3.4. Percent error in various approximate methods in Q_{ext} for a homogeneous sphere of refractive index 1.05.

x	ρ_{EA}	$\rho_{EA}(m^2-1)/4$	EA	FCI	ADA	FCII
1.0	0.1025	2.63×10^{-3}	-155.26	-155.26	-143.03	-154.48
3.0	0.3075	7.88×10^{-3}	-23.41	-23.44	-17.48	-23.02
5.0	0.5125	1.31×10^{-2}	-9.47	-9.55	-4.32	-9.23
10.0	1.025	2.63×10^{-2}	-2.47	-2.77	2.19	-2.42
20.0	2.05	5.25×10^{-2}	0.34	-0.88	4.04	-0.48
30.0	3.075	7.88×10^{-2}	2.15	-0.62	4.36	-0.15
40.0	4.10	1.05×10^{-1}	4.33	-0.60	4.42	-0.09
50.0	5.125	1.31×10^{-1}	6.85	-0.66	4.34	-0.17
60.0	6.15	1.58×10^{-1}	8.65	-0.66	4.13	-0.34
70.0	7.175	1.84×10^{-1}	6.69	-0.65	3.79	-0.75
80.0	8.20	2.10×10^{-1}	0.89	-0.87	3.57	-1.00
90.0	9.225	2.35×10^{-1}	-1.19	-1.33	3.64	-0.90
100.0	10.25	2.63×10^{-1}	-1.22	-1.68	3.71	-0.83

From Sharma and Somerford (1999).

(ii) The condition $x \gg 1$ is essentially a consequence of the requirement that the refractive index varies slowly over a wavelength. For a homogeneous sphere, the refractive index is constant. Thus, increasing x need not result in increased accuracy for a homogeneous particle. For an inhomogeneous particle, however, increase in x should result in improvement of results.

(iii) The requirement for slow variation of the refractive index is not satisfied at the boundary where there is a sharp cut-off.

The validity of the EA has been examined for the extinction efficiency factor as well (Sharma and Somerford, 1989; Sharma, 1993). It is found that the EA no longer performs better than the ADA in the domain $x \geq 1$ and $\rho_{EA} \leq 4$. Table 3.4 shows the percent errors in extinction efficiency for a nonabsorbing homogeneous sphere of $n = 1.05$ in various approximate methods. It is clear from the numbers in Tables 3.3 and 3.4 that all approximations work better for Q_{ext} in comparison with $i(0)$ when $x \geq 5.0$. For smaller values of x, however, the performance of the EA as well as the ADA is not good. This is because – for small values of x – neither the EA nor the ADA correctly reproduce the real part of the forward scattering function and, in fact, it is this real part of the scattering function which is directly proportional to the extinction efficiency factor via the extinction theorem (3.7).

The effect of a small absorptive part ($n' = 0.1$) on $i(0)$ and Q_{ext}, respectively, has been examined by Chen (1988). EA as well as ADA predictions have been compared with Mie theory results. It has been demonstrated that – as a consequence of the presence of the absorption – forward scattering results are better approximated in the EA as well as in the ADA. The oscillations noted in percent errors for nonabsorbing spheres now begin to die down. It is also noted that the EA cannot reproduce the ripple structure which is due to the partial wave resonances. However, it may be mentioned here that the behavior of the extinction curves averaged over a size distribution is reproduced quite well.

The inclusion of corrections improve the results considerably. The modified approximations are found to work extremely well in the domains $x \geq 5.0$ and $\rho_{EA}|n^2 - 1|/4 < 1$ (Sharma, 1993). It is clear from Tables 3.3 and 3.4 that a significant improvement is achieved for $i(0)$ as well as for Q_{ext}. In particular, the outcome of $S(\theta)_{FCII}$ is found to be very good. Results do not improve, rather they deteriorate, for $x \leq 5$. This, in fact, is the region where the effect of neglect – of vector nature – is significant. Hence, correcting only for the EA does not gives rise to improvement in approximate scattered intensity. Unfortunately, it is difficult to correct the EA systematically in the scalar approximation. Nevertheless, the difficulties associated with the region $x \leq 5$ may be removed for $i(0)$ for nonabsorbing homogeneous spheres by introducing an empirical multiplicative factor. Sharma et al. (1982) define a modified amplitude as:

$$i(0)_{MFCI} = A(x,n)i(0)_{FCI},$$

where

$$A(x,n) = 1 - \frac{4(n-1)}{x(n^2+1)} + \frac{(n-1)}{x^2(n^2+2)}.$$

This multiplicative factor is not unique and has been arrived at by noting that $A(x,n)$ should be close to $i(0)_{exact}/i(0)_{scalar}$, thus taking into account the effect of vector corrections. Applying this expression gives $A(x) = 0.787, 0.919$ and 0.975 for $x = 1, 3$ and 10, respectively, for $m = 1.15$. These values are essentially identical to those given in Table 3.2. The results presented in Table 3.3 show that this procedure dramatically improves the FCI results in the domain $x \leq 5$.

In the context of potential scattering, it was noted in Chapter 2 that the EA is expected to be a good approximation at all angles if $\chi(b)_{EA}$ is singular at $b = 0$ (see Section 2.8). For a homogeneous sphere, $\chi(b)_{EA}$ is analytic at $b = 0$. Hence, the approximation is not expected to be valid at large scattering angles. Not many numerical studies into angular variation of $i(\theta)$ have been performed. Whilst Chen (1988, 1989) and Chen and Smith (1992) have compared the angular scattering patterns of the EA, the ADA and the GEA with exact results, Perrin and Chiappetta (1985) and Sharma et al. (1988a) have examined the EP and the EA against exact results. The studies were conducted for dielectric particles. The following conclusions emerge from these studies:

(i) The EA works to within a 25% error in the domain $x \geq 1$, $1 \leq n \leq 1.2$ and $\theta \leq 10.0°$.
(ii) The GEA method greatly improves the EA results. More importantly, it appears to work very well for n as large as 4.0. However, success is only for scattering angles up to $\sim 5°$. Its improved variant, MGEA, is found to work well for the scattering of light with perpendicular polarization. It predicts positions of minima and maxima well for θ up to $60°$, $x \geq 5.0$, $n \leq 4$ and $n' \leq 0.5$.
(iii) The simplified version of the GEA, given in (3.84), is found to work just as well as the GEA itself for $x > 10$.

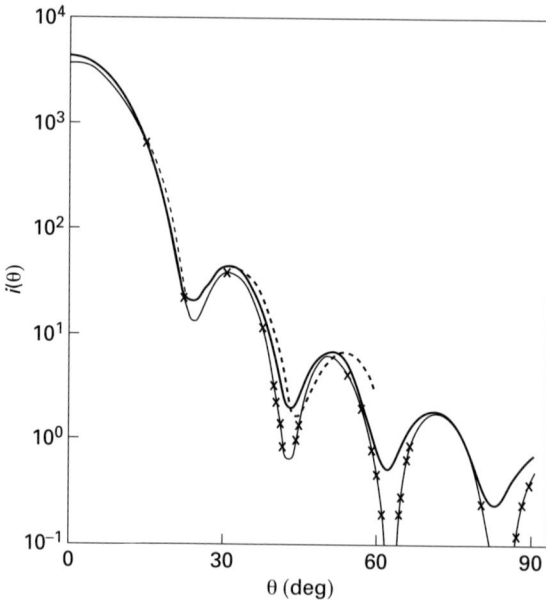

Figure 3.3. Scattered intensity $i(\theta)$ as a function of θ for a perfect homogeneous sphere with $n = 1.10$ and $x = 10.0$. Solid line: Mie theory; dashed line: EA; crosses: EP.
From Sharma et al. (1988a).

(iv) For a nonabsorbing dielectric sphere the EA and the EP agree well with the exact results for small-angle scattering. The two approximations for scattering pattern are very similar in the forward scattering lobe but the EA has some edge over the EP. Outside the forward lobe the positions of maxima and minima are determined more accurately in the EP, but the errors in the value of intensity at the positions of minima is much greater in this approximation than those obtained when the EA is used. These points are clearly illustrated in Figure 3.3.

(v) As the imaginary part of the refractive index increases, the oscillations in the scattering pattern at large angles decrease. The accuracy of the EP then increases and the EP then appears to qualify as an all-angle approximation.

The scattered intensity $|S(\theta)_{EP}^{mod}|^2$ for absorbing particles has been compared with Mie calculations over a range of particle sizes and refractive indexes. A numerical comparison between $|S(\theta)_{EP}^{mod}|^2$ and $|S(\theta)_{EP}|^2$ with Mie intensity has been performed by Bourrely et al. (1996) for $x = 20.0$ and $m = 1.5 + i0.1$. The modified formula is in better agreement with Mie theory than the original eikonal formula for scattering angles greater than $40°$. The model is valid for $x \geq 10.0$.

The two-wave WKBA and the single-wave WKBA have not been compared with the EA or any of its variants. Their validity has been examined against exact results by Klett and Sutherland (1992) and by Jones et al. (1996). Error contour charts have been plotted by Jones et al. (1996) for spheres in the domain $1.0 < n < 1.5$ and $0 < x < 20.0$. The results showed that the two-wave WKBA was superior to both the RGDA and the single-wave WKBA models. However, this superiority was noted to be very limited, especially at angles greater than $\pi/2$. At these angles, the errors almost always exceeded 10%. The introduction of absorption damps the wave inside the particle and as a result the effect of the second wave diminishes. However, some improvement in both single- and two-wave WKBA models has been noted for particles of size $x = 1 - 2$ when absorption is not too strong ($m = 1.33 + i0.1$) (Klett and Sutherland, 1992).

The extinction efficiency factor with edge effects given by (3.44) has been compared with the Mie extinction efficiency factor and the extinction efficiency factor in the ADA. Comparisons reveal that edge effects greatly improve the estimate of Q_{ext} (Ackerman and Stephens, 1987).

3.4.6 One-dimensional models

Further insight into the validity conditions of the EA may be gained by examining one-dimensional models. Clearly, a one-dimensional model constitutes a tremendous simplification of what one encounters in reality, but it has the compensating advantage of being exactly soluble. For this reason one-dimensional models have been used for a long time to study the validity domain of the Glauber multiple scattering series and the eikonal-type approximations in the context of potential scattering (Tobocman and Pauli, 1972; Chen, 1974; Banerjee *et al.*, 1975) and in examining optical scattering approximations (Alvarez-Estrada and Calvo, 1981; Sharma *et al.*, 1988b; Lin and Fiddy, 1992) and have successfully served as a useful guide to more realistic three-dimensional calculations.

For a one-dimensional homogeneous scatterer the transmission amplitude is (Sharma *et al.*, 1988b):

$$f^+ = \frac{4m}{(m+1)^2}\left[1 - \frac{(m-1)^2}{(m+1)^2}e^{4ixm}\right]e^{2ix(m-1)},$$

$$= \frac{4m}{(m+1)^2}\left[1 - \frac{(m-1)^2}{(m+1)^2}e^{4ixm}\right]f^+_{ADA}. \quad (3.99)$$

For the sake of simplicity let us take n' to be 0. Then, it can easily be seen from (3.99) that $|f^+_{ADA}|^2$ tends to $|f^+|^2$ as $n \to 1$. However, this transition occurs in an oscillating manner because of the presence of an exponential term. But if the value of nx is kept fixed, the approximation should improve continuously as $n \to 1$. This is also true for any fixed value of n'.

The equation (3.99) may also be written as:

$$f^+ = \frac{e^{-ix(m^2-1)^2}e^{ix(m^2-1)}}{1 + (m-1)^2[(1-e^{4imx})/4m]},$$

$$= \frac{e^{-ix(m^2-1)^2}f^+_{EA}}{1 + (m-1)^2[(1-e^{4imx})/4m]}. \quad (3.100)$$

In relation (3.100) the same arguments apply for the validity of the EA as those for the validity of the ADA in (3.99).

Next, let us consider a continuously varying refractive index profile of the form:

$$n^2(z) = 1 + \frac{\Lambda(\Lambda-1)}{x^2}\operatorname{sech}^2\frac{kz}{x}, \qquad (3.101)$$

where

$$\Lambda(\Lambda-1)/x^2 \equiv (n^2(0)-1) = (n^2-1),$$

and $x/k = a$ determines the characteristic length over which $n^2(z) - 1$ drops by an order of magnitude – that is, it essentially determines the size of the scatterer. The transmission amplitude for this problem can be written as (Flugge, 1971):

$$f^+ = \tfrac{1}{2}\left[e^{2i\phi_e} - e^{2i\phi_0}\right],$$

where

$$\phi_e = \arg\left[\sqrt{\pi}\frac{\Gamma(ix)\exp(ix\ln 2)}{\Gamma\left(\frac{\Lambda}{2}+\frac{ix}{2}\right)\Gamma\left(\frac{1-\Lambda}{2}+\frac{ix}{2}\right)}\right],$$

and

$$\phi_0 = \arg\left[\frac{\sqrt{\pi}}{2}\frac{\Gamma(ix)\exp(ix\ln 2)}{\Gamma\left(\frac{1+\Lambda}{2}+\frac{ix}{2}\right)\Gamma\left(1-\frac{\Lambda}{2}+\frac{ix}{2}\right)}\right].$$

To examine the condition for which f_{EA}^+ constitutes a close approximation to the exact f^+, ϕ_e and ϕ_0 may be expanded in powers of (n^2-1) and $1/x$. To this end f^+ can be written correctly up to order $(n^2-1)^2$ and $1/x$ as (Sharma et al., 1988b):

$$f^+ = \exp\left[i\left(\rho_{EA} - \frac{\rho_{EA}^2}{6x} + \frac{\rho_{EA}^2}{30x^2}\right)\right]. \qquad (3.102)$$

The first term in the exponential is nothing more than the eikonal phase. Clearly, the EA improves as $x \to \infty$ for fixed ρ_{EA}. That is, the EA needs to be looked upon as an $n \to 1$ approximation for continuously varying index profiles too. But, in addition, it should be noted from (3.102) that as x increases for a fixed value of Λ the EA also improves. That is, for an inhomogeneous scatterer the EA must be looked upon as a $x \gg 1$ approximation.

In another one-dimensional study, Alvarez-Estrada and Calvo (1981) have considered the transmission amplitude for an inhomogeneous dielectric slab where the dielectric permittivity ε is assumed to be dependent on frequency ω at each point z through the Kramer–Kronig dispersion relation. The authors find that the EA can be regarded as a $k \to \infty$ approximation here because the permittivity:

$$\varepsilon(z,\omega) = 1 - \frac{\varepsilon_1(z)}{\omega^2}$$

approaches unity in this limit. But, if $\varepsilon(z)$ is independent of ω, then the EA can no

Sec. 3.4] Scattering by a homogeneous sphere 47

longer be looked upon as a $k \to \infty$ approximation. This result is in agreement with the outcome of the above study by Sharma et al. (1988b).

3.4.7 Backscattering

Although the EA has been derived as a near forward scattering approximation, surprisingly it can also serve as a useful basis to describe back scattering if conditions (3.22) and (3.23) are satisfied. This has been achieved in theories in which large-angle scattering is a consequence only of a few hard scatterings (scatterings as viewed in the Born series) rather than the accumulation of a number of small-angle scatterings. The Saxon and Schiff approximation (Saxon and Schiff, 1957) assumes that backscattering is due to a single hard scattering event. The collisions before and after hard scattering were considered to be soft and, hence, were treated in the EA. The resulting Born series can then be summed to yield the following closed form expression for a homogeneous spherical scatterer:

$$S(\pi)_{SS} = -i\frac{(1-m^2)x}{8}\left[e^{2ixm^2} + e^{-2ix}\right]. \qquad (3.103)$$

The subscript SS stands for Saxon and Schiff. The scattering process is depicted schematically in Figure 3.4a. There are two reflections, both at impact parameter 0: one at the front surface and the other at the back surface. A comparison of $i(\pi)_{SS}$ with the exact result for a nonabsorbing sphere of $n = 1.05$ has been shown in Figure 3.5a for x ranging from 5.0 to 50.0. Approximate results are in good agreement with exact results except at those values of size where scattered intensity has a minimum. Interestingly, the positions of minima are reproduced quite accurately.

The contributions of two and three hard scattering events – shown in Figures 3.4b and 3.4c, respectively – have been calculated for a three-dimensional square well potential by Reading and Bassichis (1972). In the context of the scattering of light by a sphere these additional contributions have been calculated by Sharma and Somerford (1994). These are found to be:

$$S(\pi)^I_{SS} = \frac{x(m^2-1)^2}{8}\sqrt{\frac{x\pi}{\sqrt{2}}} e^{2ix\sqrt{2}+(3ix/\sqrt{2})}(m^2-1) - \frac{i\pi}{4}, \qquad (3.104)$$

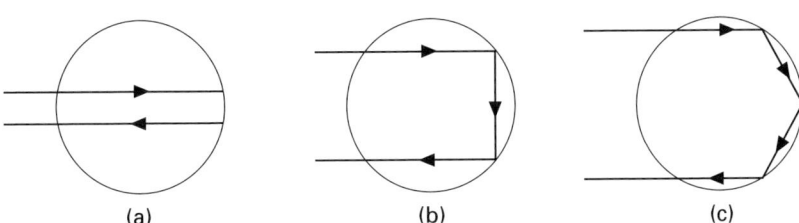

Figure 3.4. Description of backscattering process of a wave by (a) single hard scattering, (b) double-scattering and (c) triple-scattering.
From Sharma and Somerford (1999).

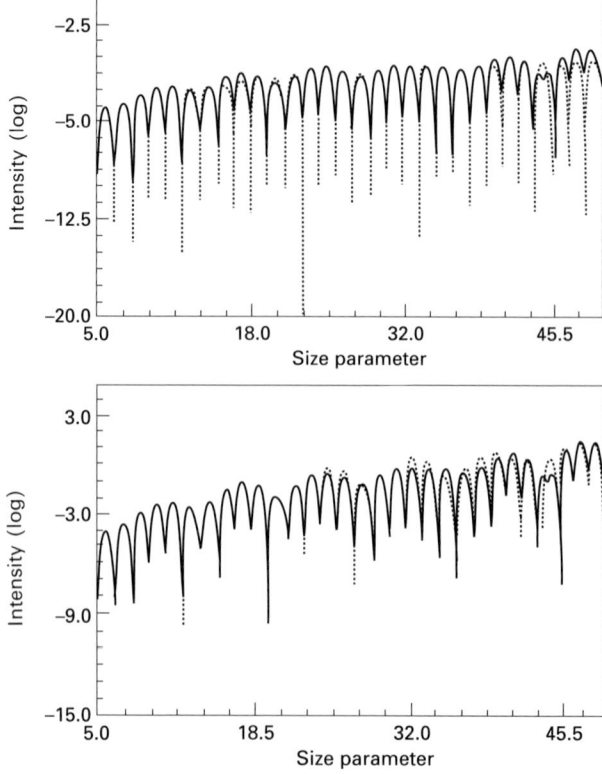

Figure 3.5. (a) Comparison of $\log i(\pi)_{SS}$ (dotted line) with $i(\pi)_{MIE}$ (solid line) for $m = 1.05$. (b) Comparison of $\log |S(\pi)_{SS} + S(\pi)^I_{SS}|^2$ (dotted line) with $i(\pi)_{MIE}$ (solid line) for $m = 1.05$. From Sharma and Somerford (1994).

and

$$S(\pi)^{II}_{SS} = -ix(m^2 - 1)^3 \sqrt{6\pi x}\, e^{3ix + 2ix(m^2 - 1) - (i\pi/4)}. \tag{3.105}$$

If the medium is highly absorbing, it is clear from (3.103) that the main contribution to scattered intensity comes from the first reflection. In contrast, for a nonabsorbing dielectric sphere:

$$\frac{|S(\pi)_{SS}|^2}{|S(\pi)^I_{SS}|^2} = x(n^2 - 1)^2,$$

and the contribution from two hard scattering events becomes important and may even dominate if $x > 1/|n^2 - 1|^2$. Figure 3.5b shows the effect of two hard scattering events. The deep minima predicted by $i(\pi)_{SS}$ now get filled up to give values at these points that are closer to exact values.

The scattered intensity corresponding to (3.103) for a nonabsorbing sphere is:

$$i(\pi)_{SS} = \frac{x^2(n^2 - 1)^2}{16} \cos^2[x(n^2 + 1)]. \tag{3.106}$$

Table 3.5. Comparison of the average separation between two successive minima with prediction of (3.107).

m	1.01	1.02	1.03	1.04	1.05	1.06	1.07	1.08	1.09	1.10
Mie	1.562	1.544	1.526	1.508	1.490	1.474	1.458	1.426	1.418	1.410
Appx	1.555	1.540	1.524	1.509	1.494	1.479	1.465	1.450	1.435	1.421

From Sharma and Somerford (1994).

The positions of minima in the scattering pattern are then given by the relation $i(\pi)_{SS} = 0$, leading to the condition:

$$x(n^2 + 1) = p\pi,$$

where p is an integer. The separation between two successive minima for a given n is thus related to the refractive index through the relation:

$$\Delta x = \frac{\pi}{n^2 + 1}. \tag{3.107}$$

Table 3.5 shows a comparison of true separation between minima and those predicted by (3.107) for various values of n. The results are based on the average separation obtained from the first 11 minima beginning at $x = 5.0$. It may be seen that predicted separation is in good agreement with actual separation. Equation (3.107), therefore, is a potentially useful relation for diagnostic purposes.

Another backscattering formula based on a single hard scattering event has been given by Chen and Hoock (1975). This differs from the Saxon and Schiff derivation in the way it takes into account re-scattering effects (Chen, 1984). It is found to give an improvement over the Saxon and Schiff results, but is less accurate than $S(\pi)_{SS} + S(\pi)_{SS}^I$ (Sharma and Somerford, 1994).

The backscattering formulas in single- and two-wave WKBAs have been obtained by Klett and Sutherland (1992). The backscatter efficiency, defined as:

$$Q_{back} = \frac{1}{x^2}\left|\sum_l (2l+1)(-1)^l(a_l - b_l)\right|^2,$$

can be expressed as:

$$Q_{back} = \frac{4}{\pi}x^2\left|\frac{m-1}{m+1}\right|^2 |I_1 + \exp(i\varrho_1)I_2|^2, \tag{3.108}$$

in the two-wave WKB approximation, where:

$$I_1 = \frac{1}{2}\left[\frac{i}{\varrho_1^2} - (i+\varrho_1)\frac{e^{i\varrho_1}}{\varrho_1^2} - \frac{i}{\varrho_2^2} + (i-\varrho_2)\frac{e^{-i\varrho_2}}{\varrho_2^2}\right],$$

and

$$I_2 = \frac{i}{2}\left[(i\varrho_3 - 1)\frac{e^{i\varrho_3}}{\varrho_3^2} + \frac{1}{\varrho_3^2} + \frac{1}{2}\right],$$

with

$$\varrho_1 = 2xm, \quad \varrho_2 = 2x \quad \text{and} \quad \varrho_3 = 2x(m-1).$$

Numerical comparisons with exact results show that for the moderately soft case $[m = (1.33, 0)]$, equation (3.108) agrees with the exact values to within an order of magnitude. A substantial improvement over the original WKBA is achieved. However, for the strong scattering case $[m = (3, 0)]$, improvement is only marginal.

When substantial absorption is present, the exact backscatter converges rapidly to a limiting value. The single- and two-wave WKBA backscatter also converge, but the limiting value is too low in comparison with the exact limiting value. Nevertheless, it is found that the results can be improved substantially by scaling the results by a factor:

$$\left[1 + [1 - \exp(-n'x)](2\sqrt{\pi} - 1)\right]^2.$$

The fine structure of the backscatter however, cannot be improved. Overall accuracy improves even for hard scattering.

3.4.8 Vector description

Attempts to incorporate the vector nature of light in the description of the EA to get access to polarization have been made by Chiappetta (1980), Perrin and Lamy (1983, 1986) and Bourrely et al. (1991).

The initial approaches to achieve this consist in writing the scattering function as a sum of diffracted and reflected parts (Chiappetta, 1980; Perrin and Lamy, 1983). The role of the EA in these approaches is limited to modeling the diffracted part. A shape-dependent reflective model is adapted for the reflected part. Therefore, polarization arises entirely due to reflection. This model has been applied to various particles that are of interest in astrophysics, and it has been found that the model gives reasonably correct results as long as $n' < 0.05$.

A more direct approach for a spherical particle stems from the identification of a scalar eikonal solution with the perpendicular component $S_1(\theta)$ of the scattered field. It can then be shown (Perrin and Lamy, 1986) that the parallel component can be obtained for large particles by multiplying $S_1(\theta)$ by the ratio of the reflectivities which are obtained from single and double reflections (Wolf, 1980, 1981). Polarization arises due to reflectivities in this approach too.

The separate eikonalization of the two Mie amplitudes is the most logical way to proceed whenever possible. That is, whenever exact solutions are available. For a sphere this has been performed by Bourrely et al. (1991). The method is essentially the same as given in Section 3.4.2. Simple formulas valid for absorbing particles have been obtained for vector eikonal functions. These read as:

$$S_1(\theta) = S_1^{diff}(\theta) H(\theta_{max} - \theta) + k \int_0^a db \left[e^{2i\alpha(b)} \frac{\cos\theta}{\sin\theta} J_1(kb\sin\theta) \right.$$

$$\left. + e^{2i\beta(b)} \left[kb J_0(kb\sin\theta) - \frac{\cos^2\theta}{\sin\theta} J_1(kb\sin\theta) \right] \right], \quad (3.109)$$

where

$$S_1^{diff}(\theta) = \frac{1}{\sin\theta}\left[(1-\cos\theta)\frac{\cos\theta}{\sin\theta}[1-J_0(x\sin\theta)] + xJ_1(x\sin\theta)\right],$$

is the diffractive component which is 0 for $\theta > \theta_{max} = 180/x$ radians, and $\alpha(b)$ and $\beta(b)$ are as defined in Section 3.4.2. The scattering function for the other polarization component $S_2(\theta)$ is obtained from (3.109) by permuting $\alpha(b)$ and $\beta(b)$.

The validity of this formalism has been investigated numerically against Mie predictions. Comparisons were made for the scattered intensity and the degree of polarization for a homogeneous sphere with $a = 500\,\mu\text{m}$, $m = 1.10 + i0.01$ and $\lambda = 0.6283\,\mu\text{m}$. It was observed that the degree of polarization is in satisfactory agreement with Mie theory for $i(\theta)$ except for $\theta > 160^0$, where the eikonal model predicts oscillations not present in Mie calculations. The degree of polarization was also found to be in perfect agreement with Mie theory predictions for $\theta > \theta_{max}$. This is better for particles of larger sizes. Comparisons have also been performed for $a = 500\,\mu\text{m}$ and a particle of greater refractive index $m = 1.9 + i0.01$. The results are very similar to those obtained for a soft scatterer. The possibility of applying the above approach to nonspherical particles has also been examined (Bourrley *et al.*, 1991). For an ellipsoidal particle, it is found that this approach can be used with reasonable accuracy if $(r - 1)$ is not very large, r being the ratio of major axis to minor axis.

3.5 INFINITELY LONG CYLINDER

The scattering of light by an infinitely long, homogeneous, circular cylinder is another instance where the EA has been examined in detail. This model of scattering lends itself to many practical situations and, perhaps, is the second most widely employed model in light scattering applications. The exact solution of the problem was first obtained by Rayleigh (1881) for perpendicular incidence. Wait (1955) solved the problem for oblique incidence and Adey (1956) extended the solution to a concentric cylinder. Hart and Montroll (1951) and Montroll and Hart (1951) considered the scalar wave analogy to arbitrary incidence and, subsequently, Montroll and Greenberg (1952) extended this to soft inhomogeneous cylinders. Yeh (1963) developed the solution for elliptic cylinders.

3.5.1 Normal incidence

A plane electromagnetic wave incident normally on an infinite long cylinder can be considered a superposition of transverse electric and transverse magnetic waves. If the electric vector is perpendicular to the axis of the cylinder, the wave is termed "transverse electric". On the other hand, if the magnetic vector is perpendicular to the cylinder axis the wave is termed "transverse magnetic". In either case, both the electric as well as the magnetic vector are perpendicular to the direction of propagation. The scattering geometry is shown in Figure 3.6. The plane (x, y) is the scattering

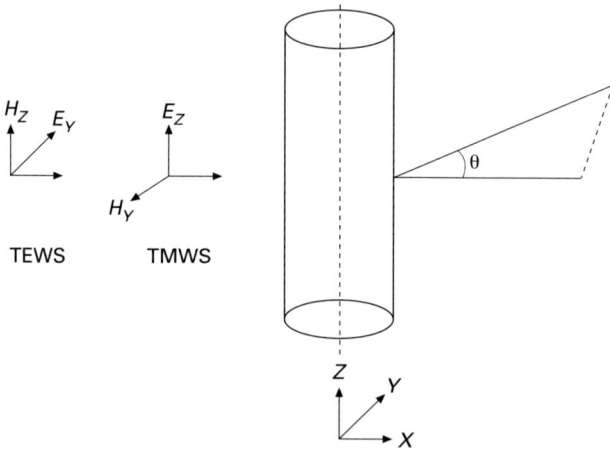

Figure 3.6. Scattering geometry for an infinitely long cylinder.

plane and the direction of the incident beam is taken to be the x-axis. For TMWS, it can easily be shown that Maxwell's equations reduce to the following two-dimensional equation:

$$[\nabla_\perp^2 + k^2 m^2(x,y)]E(x,y) = 0, \quad (3.110)$$

without any approximation. In equation (3.110), $m(x,y)$ is the complex refractive index of the cylinder relative to the surrounding medium and:

$$\nabla_\perp^2 = \frac{\partial^2}{\partial x^2} + \frac{\partial^2}{\partial y^2}.$$

Maxwell's equations take the form (3.110) for TEWS too, but only approximately. The requirement is that $ka \gg 1$.

The differential equation (3.110) can also be cast in the following integral equation form:

$$E(\mathbf{b}) = e^{i\mathbf{k}\cdot\mathbf{b}} - \frac{k^2}{4i} \int d^2\mathbf{b}' \, H_0^{(1)}(k|\mathbf{b}-\mathbf{b}'|)[m^2(\mathbf{b}') - 1]E(\mathbf{b}'), \quad (3.111)$$

where $\mathbf{b}$ is a two dimensional vector in the (x,y) plane and

$$H_0^{(1)}(k|\mathbf{b}-\mathbf{b}'|) = \frac{1}{i\pi^2} \int d^2\mathbf{p} \frac{e^{i\mathbf{p}\cdot(\mathbf{b}-\mathbf{b}')}}{p^2 - k^2 - i\epsilon}, \quad (3.112)$$

is the two-dimensional Green's function or propagator. For $|\mathbf{b}| \to \infty$, one can write the electric field as:

$$E(\mathbf{b}) \sim e^{i\mathbf{k}\cdot\mathbf{b}} + \frac{e^{ik|\mathbf{b}|}}{|\mathbf{b}|^{1/2}} T(\mathbf{k}_i, \mathbf{k}_f),$$

and from (3.111) the scattering function $T(\mathbf{k}_i, \mathbf{k}_f)$ can be identified with:

$$T(\mathbf{k}_i, \mathbf{k}_f) = \frac{-ik^2}{4} \int d\mathbf{b}\, e^{-i\mathbf{k}_f \cdot \mathbf{b}} [m^2(\mathbf{b}) - 1] E_z(\mathbf{b}). \tag{3.113}$$

The method to invoke the EA for an infinitely long cylinder is similar to that for a sphere. A trial solution:

$$E(\mathbf{b}) = e^{ikx} \phi(\mathbf{b}), \tag{3.114}$$

is chosen and substituted in (3.113). This results in the following equation:

$$\left[\nabla_\perp^2 + 2ik \frac{\partial}{\partial x}\right] \phi(\mathbf{b}) = -k^2 [m^2(\mathbf{b}) - 1] \phi(\mathbf{b}).$$

If the term containing $\nabla_\perp^2$ is ignored in the above equation, the resulting approximate equation is:

$$2ik \frac{\partial}{\partial x} \phi(\mathbf{b}) = -k^2 [m^2(\mathbf{b}) - 1] \phi(\mathbf{b}),$$

which with boundary condition $\phi(\mathbf{b}) = 1$ at $x = -\infty$ yields the solution:

$$E(x, y) = \exp\left[ikx + \frac{ik}{2} \int_{-\sqrt{a^2-y^2}}^{x} (m^2(x', y) - 1)\, dx'\right], \tag{3.115}$$

where a is the radius of the cylinder. This equation has interpretation identical to (2.24) as given in Section 2.2.4. Theoretically, the inequalities:

$$|m^2(\mathbf{b}) - 1| \ll 1 \quad \text{and} \quad x \gg 1$$

govern the validity domain of the EA here too. The solution (3.115), when substituted in (3.113), gives:

$$T(\theta)_{EA} = -\frac{k}{2} \int dx\, dy\, e^{i\mathbf{q}\cdot\mathbf{b}} \frac{\partial}{\partial x}\left[\exp\left((ik/2) \int_{-\sqrt{a^2-y^2}}^{x} [m^2(x', y) - 1]\, dx'\right)\right],$$

where $\mathbf{q} = \mathbf{k}_i - \mathbf{k}_f$. Further, for small-angle scattering, $\mathbf{q}$ is nearly perpendicular to the x-axis and one may approximate $\mathbf{q}\cdot\mathbf{b}$ by qy, where $q = -k \sin \theta$. The EA then gives:

$$T(\theta)_{EA} = -\frac{k}{2} \int_{-a}^{a} dy\, e^{iqy} [e^{i\chi(y)_{EA}} - 1], \tag{3.116}$$

where

$$\chi(y)_{EA} = \frac{k}{2} \int_{-\sqrt{a^2-y^2}}^{\sqrt{a^2-y^2}} [m^2(x', y) - 1]\, dx'. \tag{3.117}$$

The real part of (3.117) determines the phase shift suffered by the ray in traveling undeviated across the cylinder and the imaginary part decides the absorption.

Homogeneous cylinder

For a homogeneous cylinder the relative refractive index $m(\mathbf{r}) = m$. For this case:

$$\chi(y)_{EA} = 2ik(m^2 - 1)\sqrt{a^2 - y^2}$$

and the scattering function (3.116) converts to:

$$T(\theta)_{EA} = -\frac{k}{2}\int_{-a}^{a} dy\, e^{iqy}\left[e^{ik(m^2-1)\sqrt{a^2-y^2}} - 1\right]. \tag{3.118}$$

Making a change of variable $y = a\sin\gamma$, the above equation can also be cast in the following alternative form:

$$T(\theta)_{EA} = x\int_0^{\pi/2} d\gamma \cos\gamma \cos(z\sin\gamma)[1 - \exp(w^*\cos\gamma)], \tag{3.119}$$

where

$$z = x\sin\theta \quad \text{and} \quad w^* = i\rho_{EA}^* = i\rho_{EA}(1 + i\tan\beta).$$

For a completely absorbing cylinder only the first term on the right-hand side of (3.119) contributes. The resulting integration can be evaluated to give:

$$T(\theta)_{EA} - \frac{\sin(2x\sin\theta)}{2\sin\theta},$$

which is nothing more than the scattering function for an infinitely long cylinder in the Fraunhofer diffraction approximation.

For a thin cylinder, such that $x|m^2 - 1| \ll 1$, one may approximate:

$$\exp(w^*\cos\gamma) \approx 1 + w^*\cos\gamma.$$

This simplification converts (3.119) to the following expression:

$$T(\theta)_{EA} \approx -w^* x \int_0^{\pi/2} d\gamma \cos^2\gamma \cos(z\sin\gamma) = -\frac{\pi x w^*}{2z} J_1(z), \tag{3.120}$$

which may be recognized as the scattering function in the RGDA for the TMWS case. This implies that the EA applies to the entire x region despite the original premise $x \gg 1$ for the TMWS case. The RGDA for the TEWS case differs from its TMWS case only by a multiplicative factor $\cos\theta$. That is:

$$T(\theta)_{RGDA}^{TEWS} = \cos\theta\, T(\theta)_{RGDA}^{TMWS}.$$

Thus, for the TEWS case the EA reduces to the RGDA only for forward scattering function. For non-forward scattering it needs to be multiplied by a factor $\cos\theta$.

For forward scattering, the integration in (3.119) can be evaluated analytically to yield:

$$T(0)_{EA} = \frac{x\pi}{2} L(w^*), \tag{3.121}$$

where

$$L(w^*) = I_1(w^*) - L_1(w^*),$$

with $I_1(\omega^*)$ as the modified Bessel function and $L_1(\omega^*)$ as the modified Struve function. For a nonabsorbing cylinder, equation (3.121) simplifies to:

$$T(0)_{EA} = \frac{\pi x}{2}[iJ_1(\rho_{EA}) + S_1(\rho_{EA})], \qquad (3.122)$$

where $J_1(\rho_{EA})$ and $S_1(\rho_{EA})$ are Bessel and Struve functions of order 1. The extinction efficiency factor Q_{ext} which is related to $T(0)$ via the optical theorem is:

$$Q_{ext} = \frac{2}{x}\text{Re }T(0), \qquad (3.123)$$

gives

$$Q_{EA}^{ext} = \pi \text{ Re }[L(\omega^*)], \qquad (3.124)$$

as the extinction efficiency factor in the EA for an infinitely long cylinder.

For non-forward scattering, the imaginary part of the scattering function can be evaluated analytically for a nonabsorbing cylinder to yield:

$$\text{Im }T(\theta)_{EA} = -\frac{i\pi x \rho_{EA}}{2y}J_1(y). \qquad (3.125)$$

The integration of the real part can be achieved only by means of appropriate series expansions (Sharma et al., 1997a). For this purpose, let us write the real part of the scattering function as:

$$\text{Re }T(\theta)_{EA} = x\int_0^{\pi/2} d\gamma \cos\gamma \cos(z\sin\gamma)[1 - \cos(\rho_{EA}\cos\gamma)]. \qquad (3.126)$$

The integration of the first term on the right-hand side of (3.126) (namely, $\cos\gamma\cos(z\sin\gamma)$) is trivial and gives,

$$\text{Re }T(\theta)_{EA}^1 = x\int_0^{\pi/2} d\gamma \cos\gamma \cos(z\sin\gamma) = x\frac{\sin z}{z}. \qquad (3.127)$$

This can readily be identified as the diffraction contribution. The superscript 1 on T refers to the contribution of the first term of the integration. Integration of the remaining term in the integrand in (3.126) can be carried out by parts. This yields:

$$\text{Re }T(\theta)_{EA}^2 = -x\frac{\sin z}{z} + \frac{x\rho_{EA}}{2}\int_0^1 dt \sin\left(z[1-t^2]^{1/2}\right)\sin(\rho_{EA}t). \qquad (3.128)$$

The superscript 2 refers to the contribution of the second term of the integrand. It is interesting to note that an equal and opposite contribution of $x\sin z/z$ comes from this part which cancels the diffraction contribution of the same magnitude but opposite sign. Substituting the series expansion in powers of $z(1-t^2)^{1/2}$ for $\sin(z(1-t^2)^{1/2})$ in (3.128), the dt integration of each term in the resulting integrand can be performed separately for $z < 1$, yielding:

$$\text{Re }T(\theta)_{EA} = \frac{\pi x}{2}\sum_{k=0}^{\infty}\frac{(-1)^k}{2k!!}\left(\frac{z^2}{\rho_{EA}}\right)^k S_{k+1}(\rho_{EA}). \qquad (3.129)$$

For forward scattering, only the $k = 0$ term in (3.129) contributes, whereby one gets:

$$\operatorname{Re} T(0)_{EA} = \frac{x\pi}{2} S_1(\rho_{EA}),$$

which is indeed what was obtained in (3.122).

Instead of writing $\sin(z(1-t^2)^{1/2})$ as a series in (3.128), one may alternatively express $\sin(\rho_{EA} t)$ as a series in powers of $\rho_{EA} t$. The integration, then, leads to the following series:

$$\operatorname{Re} T(\theta)_{EA} = \frac{x\rho_{EA}^2}{2} \sqrt{\frac{\pi}{2z}} \sum_{k=0}^{\infty} \frac{\Gamma(k+1)}{(2k+1)!} \left(\frac{2\rho_{EA}^2}{z}\right)^k J_{k+3/2}(z). \quad (3.130)$$

for $\rho_{EA} < 1$. Let us now examine the series expansion (3.130) for a fixed value of ρ_{EA} in the domain $z < 1$. In terms of scattering angle, this domain is defined as $\sin\theta < 1/x$. One can then write:

$$T(\theta)_{EA}^{SA} = \frac{1}{2}\pi x \left[S_1(\rho_{EA}) - \frac{z}{\rho_{EA}} S_2(\rho_{EA}) \right] - i \frac{\pi x \rho_{EA}}{2y} J_1(y). \quad (3.131)$$

where the superscript SA refers to "small angle". Formula (3.131) is expected to be valid over the entire range of refractive index and size parameter values for which the EA is valid.

For a large diameter cylinder, such that $x \gg 1$ and $\rho_{EA} \gg 1$, one can make use of the following asymptotic expansion:

$$S_{k+1}(\rho_{EA}) \approx Y_{k+1}(\rho_{EA}) + \frac{1}{\pi}\sum_{s=0}^{p-1} \frac{\Gamma(s+\frac{1}{2})(\frac{1}{2}\rho_{EA})^{-2s+k}}{\Gamma(k+\frac{3}{2}-s)} + O(|\rho_{EA}|^{k-2p-1}), \quad (3.132)$$

where p is a natural number. A little manipulation of (3.129) with the help of (3.132) gives:

$$\operatorname{Re} T(\theta)_{EA} = \frac{\pi x \rho_{EA}}{2y} Y_1(y) + \frac{x \sin z}{z}. \quad (3.133)$$

The above formula is correct to order $1/\rho_{EA}$. Thus, one can write the sum of (3.125) and (3.133) as:

$$T(\theta)_{EA} = \frac{\pi x \rho_{EA}}{2y}(Y_1(y) - iJ_1(y)) + \frac{x \sin z}{z}, \quad (3.134)$$

where the first two terms on the right-hand side constitute the refraction contribution and the third term is the diffraction contribution. The scattered intensity, defined as:

$$i(\theta) = |T(\theta)|^2,$$

and corresponding to (3.134) is:

$$i(\theta)_{EA} = \frac{x^2 \sin^2 z}{z^2} + \frac{x^2 \pi \rho_{EA} \sin z}{yz} Y_1(y) + \frac{x^2 \pi^2 \rho_{EA}^2}{4y^2} [Y_1^2(y) + J_1^2(y)]. \quad (3.135)$$

A simplified version of the above equation:

$$i(\theta)_{EA} = \frac{x^2 \sin^2 z}{z^2} + \frac{x^2 \pi \rho_{EA} \sin z}{yz} Y_1(y) + \frac{x^2 \pi^2 \rho_{EA}^2}{2y^3}, \qquad (3.136)$$

can be obtained by noting that:

$$Y_1^2(y) + J_1^2(y) \simeq (2/\pi y)$$

in the large y limit. This formula describes refraction corrections to the Fraunhofer diffraction and has been referred to as the modified diffraction formula (MDF) (Sharma et al., 1997a).

The theory of light scattered by an infinitely long homogeneous cylinder is valid also for a normally illuminated cylinder composed of uniaxial material, where the axis of the cylinder coincides with the optic axis. For such a cylinder, the permittivity is characterized by a dyadic of the form:

$$\varepsilon = \begin{pmatrix} \varepsilon_\perp & 0 & 0 \\ 0 & \varepsilon_\perp & 0 \\ 0 & 0 & \varepsilon_\parallel \end{pmatrix}.$$

If the incident light is polarized parallel to the cylinder axis (TMWS), the cylinder scatters and absorbs light as if it were isotropic with permittivity $\varepsilon_\parallel$. On the other hand, if the incident light is polarized perpendicular to the cylinder axis (TEWS), it scatters light as if it were isotropic with permittivity $\varepsilon_\perp$.

3.5.2 Derivation from exact solutions

The exact scattering functions for transverse magnetic and transverse electric waves, respectively, are given by (see, e.g., Bohren and Huffman, 1983):

$$T(\theta)_{TMWS} = \sum_{l=-\infty}^{\infty} b_l e^{-il\theta} = \frac{1}{2} \sum_{l=-\infty}^{\infty} \left[1 - e^{2i\beta_l}\right] e^{-il\theta}, \qquad (3.137)$$

and

$$T(\theta)_{TEWS} = \sum_{l=-\infty}^{\infty} a_l e^{-il\theta} = \frac{1}{2} \sum_{l=-\infty}^{\infty} \left[1 - e^{2i\alpha_l}\right] e^{-il\theta}. \qquad (3.138)$$

The phase angles β_l and α_l and the scattering coefficients a_l and b_l are as follows:

$$\tan \beta_l = \frac{m J_l'(mx) J_l(x) - J_l(mx) J_l'(x)}{m J_l'(mx) N_l(x) - J_l(mx) N_l'(x)}, \qquad (3.139)$$

$$\tan \alpha_l = \frac{J_l'(mx) J_l(x) - m J_l(mx) J_l'(x)}{J_l'(mx) N_l(x) - m J_l(mx) N_l'(x)}, \qquad (3.140)$$

$$b_l = \frac{m J_l'(mx) J_l(x) - J_l(mx) J_l'(x)}{m J_l'(mx) H_l(x) - J_l(mx) H_l'(x)}, \qquad (3.141)$$

and

$$a_l = \frac{J_l'(mx)J_l(x) - mJ_l(mx)J_l'(x)}{J_l'(mx)H_l(x) - mJ_l(mx)H_l'(x)}, \quad (3.142)$$

where J_l, N_l and H_l are, respectively, the Bessel, Neumann and Hankel functions of first kind and of order l. Primes denote derivatives with respect to the argument.

The eikonal amplitude has been obtained from exact solutions by Sharma *et al.* (1988c). The procedure is analogous to what has been followed for a homogeneous spherical particle. Even the discussion pertaining to the relationship between the EA phase and the ADA phase remains unaltered. Hence, the derivation is not repeated here. Instead, we present an alternative derivation which helps in bringing out a better understanding of the validity domain of the EA.

For $|m^2 - 1| \ll 1$, which is one of the criteria for the validity of the EA, various Bessel and Hankel functions can be expanded using the following relations (Gradshteyn and Ryzhik, 1980):

$$J_l(mx) = m^l \sum_{k=0}^{\infty} \frac{(-1)^k}{k!} \left(\frac{\rho_{EA}^*}{2}\right)^k J_{l+k}(x).$$

This is a series in powers of ρ^*. Using this expansion it is possible to write b_l as:

$$b_l = -\frac{i\pi x}{2} \sum_{k=1}^{\infty} \left(\frac{\rho_{EA}^*}{2} \frac{1}{l!}\right)^k [J_l J_{l+k-1} - J_{l+k} J_{l-1}] \times \sum_{p=1}^{\infty} \left[\frac{i\pi x}{2}\right]^{p-1}$$

$$\times \left[\sum_{r=0}^{\infty} \frac{(-1)^r}{r!} \left(\frac{\rho_{EA}^*}{2}\right)^r (J_{l+r-1} H_l - J_{l+r} H_{l-1})\right]^{p-1}. \quad (3.143)$$

A cumbersome but straightforward calculation allows (3.143) to be cast as a series in powers of ρ_{EA}^* in the form:

$$b_l = A\rho_{EA}^* + B\rho_{EA}^{*2} + C\rho_{EA}^{*3} + \cdots, \quad (3.144)$$

where the coefficients A, B, C are functions of various orders of Bessel and Neumann functions. These coefficients can be simplified by using the asymptotic expansions for $J_l(x)$, $J_l'(x)$, $N_l(x)$ and $N_l'(x)$ for large values of index l, where the argument x is greater than l. These are given by the following expressions:

$$J_l(x) \sim \left[\frac{2}{\pi x} \frac{1}{[1 - (l^2/x^2)]^{1/2}}\right]^{1/2} \cos \Upsilon, \quad (3.145)$$

$$J_l'(x) \sim \left[\frac{2}{\pi x}\left[1 - \frac{l^2}{x^2}\right]^{1/2}\right]^{1/2} \sin \Upsilon, \quad (3.146)$$

$$N_l(x) \sim \left[\frac{2}{\pi x} \frac{1}{[1 - (l^2/x^2)]^{1/2}}\right]^{1/2} \sin \Upsilon, \quad (3.147)$$

and
$$N'_l(x) \sim \left[\frac{2}{\pi x}\left[1 - \frac{l^2}{x^2}\right]^{1/2}\right]^{1/2} \cos \Upsilon, \quad (3.148)$$

where
$$\Upsilon = (x^2 - l^2) - l\cos^{-1}(l/x) - \pi/4.$$

The terms neglected are of the relative order $(1/x)$. The coefficients $\mathcal{A}, \mathcal{B}, \mathcal{C}$ are then found to be:

$$\mathcal{A} = \sqrt{1 - \frac{l}{x^2}}; \quad \mathcal{B} = \frac{1}{4}\left[1 - \left(\frac{l}{x}\right)^2\right]; \quad \mathcal{C} = \frac{i}{12}\left[1 - \left(\frac{l}{x}\right)^2\right]^{3/2}.$$

Thus, b_l takes the form:
$$b_l = \tfrac{1}{2}\left[1 - i\delta_l - \tfrac{1}{2}(i\delta_l)^2 - \tfrac{1}{6}(i\delta_l)^3 - \cdots\right],$$

where
$$\delta_l = \rho^*_{EA}\sqrt{1 - (l/x)^2}.$$

Clearly, there is an indication that δ_l exponentiates, leading to the relation:
$$b_l = \tfrac{1}{2}[1 - e^{i\delta_l}].$$

Assuming that such an exponentiation indeed takes place, and employing the usual argument of replacing the summation by integration, one arrives at the conventional form of the eikonal scattering function. Proceeding in a similar manner one gets the same result for TEWS also. This deduction of the EA suggests that this approximation should be looked upon as an $x \to \infty$ approximation for fixed ρ^*. This limit, in turn, can also be interpreted as $(m - 1) \to 0$ for fixed ρ^*. Similar results have been derived in the framework of the EA for oriented, infinitely long cylinders (Sharma and Somerford, 1989).

3.5.3 Vector formalism

Equations (3.139) and (3.140) in the limit $\rho^*_{EA} \to 0$ can be written as (van de Hulst, 1957; Sharma et al., 1988c):

$$\tan \beta_l \approx \frac{\pi x \rho^*_{EA}}{4}\left[\left(1 - \frac{l^2}{x^2}\right)J_l^2 + J_l'^2\right], \quad (3.149)$$

and

$$\tan \alpha_l \approx \frac{\pi x \rho^*_{EA}}{4}\left[\left(1 - \frac{l^2}{x^2}\right)J_l^2 + J_l'^2 + \frac{2}{x}J_l J_l'\right]. \quad (3.150)$$

Further, since in the limit $\rho^*_{EA} \to 0$, $\tan \alpha_l$ and $\tan \beta_l$ also tend to 0, it is reasonable to assume that $\tan \alpha_l = \alpha_l$ and $\tan \beta_l = \beta_l$. The scattering functions (3.137) and (3.138) then turn into the following relations:

$$T_{TMWS}(\theta) \approx \sum_{l=-\infty}^{\infty} \left(1 - \exp\left[i\frac{\pi x \rho^*_{EA}}{2}\left(\left(1 - \frac{l^2}{x^2}\right)J_l^2 + J_l'^2\right)\right]\right), \qquad (3.151)$$

and

$$T_{TEWS}(\theta) \approx \sum_{l=-\infty}^{\infty} \left(1 - \exp\left[i\frac{\pi x \rho^*_{EA}}{2}\left(\left(1 - \frac{l^2}{x^2}\right)J_l^2 + J_l'^2 + \frac{2}{x}J_l J_l'\right)\right]\right). \qquad (3.152)$$

From the method of deduction of (3.151) and (3.152) it is evident that these equations would yield good predictions for exact results provided $\rho^*_{EA} \ll 1$. Further, in the limit $x, l \to \infty$ with $x > l$ – using equations (3.145) to (3.148) – the exponent can be easily shown to be nothing more than the eikonal phase $\chi(y)_{EA}$, and when the sum over l is converted into an integral, the amplitude functions are nothing more than the eikonal scattering function (3.118). Clearly, formulas (3.151) and (3.152) should be regarded as vector eikonal scattering functions valid for all values of x for an infinitely long cylinder. A comparison of angular scattered intensities from these formulas with those from the exact results are shown in Figures 3.7a and 3.7b, respectively, for TMWS and TEWS modes.

3.5.4 Corrections to the eikonal approximation

Corrections similar to those described in connection with the scattering of light by a homogeneous sphere have been studied for an infinitely long homogeneous dielectric cylinder by Sharma et al. (1981) and Sharma (1993). Simple and straightforward calculations lead to the following scattering function:

$$T(\theta)_{EA} = -\frac{k}{2}\int_{-a}^{a} dy\, e^{iqy}\left[e^{i(\chi_{EA}+\tau_1)} - 1\right], \qquad (3.153)$$

where

$$\chi_{EA} = k(m^2 - 1)\sqrt{a^2 - y^2}$$

and

$$\tau_1 = -\frac{x(m^2 - 1)(1 - 2\sin^2\gamma)}{4\cos\gamma}.$$

Note that – at $\gamma = \pi/2$ – the correction to the eikonal phase diverges. A simplified form which is free of this divergence problem can be obtained in the domain $x|m^2 - 1|/4 \ll 1$ by expanding the $\exp(i\tau_1)$ part in the exponential in (3.153). The resulting equation is:

$$T(\theta)_{FCI} = x\int_0^{\pi/2} d\gamma \cos\gamma \cos(z\sin\gamma)\left[1 - (1 + i\tau_1)\,e^{i\chi_{EA}}\right]. \qquad (3.154)$$

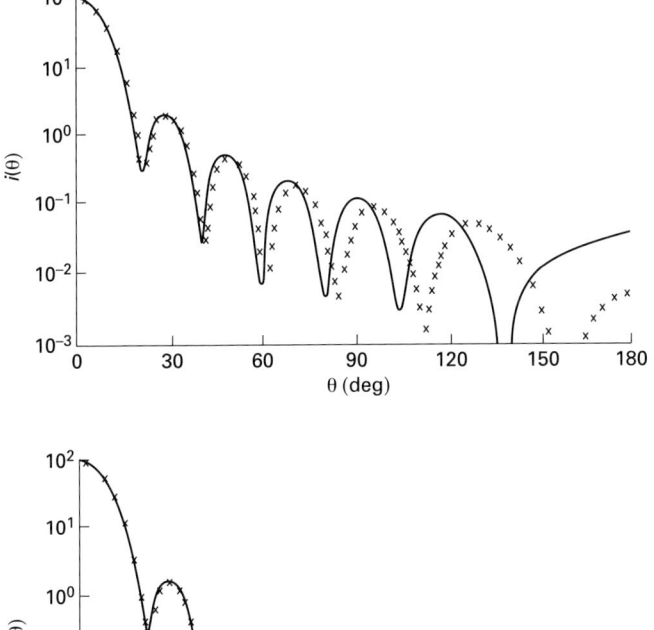

Figure 3.7. (a) Comparison of $i(\theta)_{TMWS}$ with $i(\theta)$ corresponding to (3.98a). The refractive index is $m = 1.05$ and $x = 10.0$. (b) Comparison of $i(\theta)_{TEWS}$ with $i(\theta)$ corresponding to (3.98b). The refractive index $m = 1.05$ and $x = 10.0$.

A similar correction for the ADA can be obtained as:

$$T(\theta)_{FCII} = x \int_0^{\pi/2} d\gamma \cos\gamma \cos(z\sin\gamma)\left[1 - \left(1 + i\tau_1'\right)e^{i\chi_{ADA}}\right], \tag{3.155}$$

where

$$\chi_{ADA} = 2k(m-1)\sqrt{a^2 - y^2},$$

and

$$\tau_1' = -\frac{x(m^2-1)(2-m^2)\sin^2\gamma}{4\cos\gamma}.$$

The steps involved in deduction of the above relations are identical to those for the case of a homogeneous sphere.

For forward scattering, the integrals in (3.154) and (3.155) can be evaluated analytically, leading to the following closed form formulas:

$$T(0)_{FCI} = T(0)_{EA} - \frac{\rho_{EA}^{*2}}{4}\left[\left(\frac{4\rho_{EA}^{*}}{3} {}_1F_2\left(2,\frac{3}{2},\frac{5}{2};\frac{-\rho_{EA}^{*2}}{4}\right) + \frac{i\pi}{2}S_0(\rho_{EA}^{*})\right)\right.$$
$$\left. + \frac{i\pi}{2}\left({}_1F_2\left(\frac{3}{2},\frac{1}{2},2;\frac{-\rho_{EA}^{*2}}{4}\right) - J_0(\rho_{EA}^{*})\right)\right], \quad (3.156)$$

and

$$T(0)_{FCII} = D(m)T(0)_{ADA}, \quad (3.157)$$

with

$$D(m) = [1 + 0.125 \times (m+1)^2(m-1)(2-m^2)].$$

In (3.156), $S_0(\rho_{EA}^{*})$ is the Struve function of order 0 and ${}_1F_2$ is the generalized hypergeometric function. Analogous formulas hold for (3.155) as well.

Integral (3.154) can be evaluated for arbitrary θ if the cylinder is nonabsorbing. The result is rapidly converging infinite sums. For the real and imaginary parts of the first-order corrected scattering function one obtains the following expressions (Di Marzio and Szajman, 1992):

$$\operatorname{Re} T(\theta)_{FCI} = \operatorname{Re} T(\theta)_{EA} - \frac{\rho^2}{4}\sum_{i=0}^{\infty}\frac{(-1)^i\rho^{2i+1}}{(2i+1)!}\sum_{j=0}^{i}\binom{2i+1}{2j}\left(\frac{z}{\rho}\right)^{2j}$$
$$\times \left[\sum_{k=0}^{i-j}\frac{(-1)^k}{2j+2k+1}\binom{i-j}{k}\right.$$
$$\left.\times \left[\left(\frac{i-j+1}{i-j-k+1} - \frac{2j+2k+1}{2j+2k+3}\right)\right] + \frac{(-1)^{i-j+1}}{2i+3}\right],$$

and

$$\operatorname{Im} T(\theta)_{FCI} = \operatorname{Im} T(\theta)_{EA} - \frac{\pi\rho^2}{8}\sum_{i=0}^{\infty}\frac{(-1)^i}{(2i)!}\sum_{j=0}^{i}\binom{2i}{2j}\left[\rho^{2j}z^{2(i-j)} - z^{2j}\rho^{2(i-j)}\right]$$
$$\times (-1)^k\binom{j+1}{k}\frac{(2i-2j+2k-1)!!}{(2i-2j+2k)!!},$$

where $\operatorname{Im} T(\theta)_{EA}$ and $\operatorname{Re} T(\theta)_{EA}$ are as given in Section 3.5.2. Alternatively, these may also be expressed as (Di Marzio and Szajman, 1992):

$$\operatorname{Re} T(\theta)_{EA} = x\left[\frac{\sin x}{x} - \left(\sum_{i=0}^{\infty}\frac{(-1)^i z^{2i}}{(2i)!}\sum_{j=0}^{i}\left(\frac{\rho}{z}\right)^{2j}\binom{2i}{2j}\right.\right.$$
$$\left.\left.\times \sum_{k=0}^{j}(-1)^k\binom{j}{k}\frac{1}{2i-2j+2k+1}\right)\right],$$

and

$$\text{Im } T(\theta)_{EA} = \frac{\pi}{2} x \sum_{i=0}^{\infty} \frac{(-1)^i \rho^{2i+1}}{(2i+1)!} \sum_{j=0}^{i} \left(\frac{z}{\rho}\right)^{2j} \binom{2i+1}{2j}$$

$$\times \sum_{k=0}^{j} (-1)^k \binom{j}{k} \frac{(2i-2j+2k+1)!!}{(2i-2j+2k+2)!!},$$

in the validity domain of the first-order corrected EA. Similar results can also be obtained for $T(\theta)_{FCII}$.

3.5.5 Numerical comparisons

Typical percent errors in forward scattered intensities for a homogeneous circular cylinder of relative refractive index $m = 1.05$ are shown in Tables 3.6 and 3.7, respectively, for:

$$i(0)_{TEWS} = |T(0)_{TEWS}|^2 \quad \text{and} \quad i(0)_{TMWS} = |T(0)_{TMWS}|^2.$$

The percent error is defined through the relation (3.98). It is clear from due comparison that the EA is superior to the ADA in the domain $\rho_{EA} \leq 3.5$ for TMWS. For TEWS the EA is preferable to the ADA in the domain $x \geq 3$ and $\rho \leq 3.5$. Note that the value $\rho_{EA} \approx 3.5$ corresponds roughly to the first maximum in the extinction curve for the scattering by an infinitely long circular cylinder at normal incidence. Recall that a resembling feature was also observed for the case of scattering by a homogeneous sphere.

For decreasing $|m-1|$ the accuracy of the approximation increases. It also increases with increasing x, but is modulated by an oscillation. The general trend

Table 3.6. Percent error in various approximate methods in $i(0)_{TMWS}$ for an infinitely long homogeneous cylinder of refractive index 1.05.

x	$\rho_{EA}(m^2-1)/4$	EA	FCI	ADA	FCII
0.2	5.25×10^{-4}	0.79	0.79	5.58	1.08
0.6	1.58×10^{-3}	2.62	2.62	7.31	2.89
1.0	2.63×10^{-3}	2.05	2.05	6.77	2.32
2.0	5.25×10^{-3}	0.45	0.44	5.24	0.72
3.0	7.88×10^{-3}	0.10	0.05	4.88	0.35
5.0	1.31×10^{-2}	0.11	0.01	4.83	0.30
10.0	2.63×10^{-2}	0.22	-0.17	4.69	0.14
15.0	3.94×10^{-2}	0.58	-0.32	4.59	0.04
20.0	5.25×10^{-2}	1.25	-0.37	4.59	0.04
25.0	6.57×10^{-2}	2.12	-0.46	4.58	0.03

From Sharma and Somerford (1999).

Table 3.7. Percent error in various approximate methods in $i(0)_{TEWS}$ for an infinitely long homogeneous cylinder of refractive index 1.05.

x	$\rho_{EA}(m^2-1)/4$	EA	FCI	ADA	FCII
0.2	5.25×10^{-4}	-10.04	-10.04	-4.70	-9.73
0.6	1.58×10^{-3}	-8.56	-8.56	-3.33	-8.26
1.0	2.63×10^{-3}	-7.63	-7.63	-2.45	-7.33
2.0	5.25×10^{-3}	-4.05	-4.06	-0.96	-3.77
3.0	7.88×10^{-3}	-2.97	-3.01	1.96	-2.72
5.0	1.31×10^{-2}	-1.64	-1.74	3.17	-1.45
10.0	2.63×10^{-2}	-0.67	-1.06	3.84	-0.74
15.0	3.94×10^{-2}	-0.03	-0.93	4.01	-0.57
20.0	5.25×10^{-2}	0.76	-0.87	4.12	-0.45
25.0	6.57×10^{-1}	1.70	-0.90	4.16	-0.41

From Sharma and Somerford (1999).

is akin to what was observed for a homogeneous sphere. For thin cylinders ($x < 1$), the scattered intensity in the EA is found to be a good approximation for TMWS despite the original requirement $x \gg 1$. However, performance of the EA for TEWS is comparatively poor for x close to 1. Further, a comparison of percent errors given in Tables 3.6 and 3.7 for an infinitely long cylinder with corresponding quantities in Tables 3.3 and 3.4 for the scattering by a homogeneous sphere shows that the EA and its variants are more accurate for an infinite cylinder than for a homogeneous sphere.

Comparison of the angular variation of scattered intensity in the EA, $i(\theta)_{EA}$, with exact results confirms that the EA is a small-angle approximation (Sharma and Somerford, 1982). It has been found to work fairly accurately at least up to about 30° for intermediate size scatterers for TMWS. The same is true for TEWS except at x values close to 1. As the imaginary part of the relative refractive index increases, the approximation appears to improve (Sharma and Somerford, 1983b). For very high absorption, the EA and the ADA approach the Fraunhofer diffraction formula.

The inclusion of corrections improves the results vastly. The approximations FCI and FCII are then found to work extremely well in the domain $\rho_{EA}|n^2-1|/4 < 1$. However, no advantage is achieved by making use of the more complicated form of the correction FCI than the much simpler correction FCII.

The accuracy of the extinction and absorption efficiency factors in the ADA for an infinitely long cylinder with incident, unpolarized light at normal incidence has been examined by Chýlek and Klett (1991a) for various values of n and two values of n' (0.01 and 0.1). For $\rho > 0.6$, the percent error was less than 8% for $n = 1.10$, less than 10% for $n = 1.20$ and less than 25% for $n = 1.4$. For $\rho < 0.6$, the errors decrease rapidly with increasing absorption. The maximum error decreases from 90% at $m = 1.4 + i0.01$ to approximately 7% at $m = 1.4 + i0.1$. For a homogeneous cylinder of $m = 4/3$, good agreement between the exact and the ADA extinction efficiency factors has been noted by Stephens (1984).

3.5.6 Oblique incidence

Let us denote by Ψ the angle that the wave vector of the incident plane wave makes with the z-axis of the cylinder. The phase shift suffered by any ray in traveling undeviated through the medium is now greater because the ray now travels a greater distance in the scattering medium. The new distance is nothing more than the distance it would travel if the incidence was perpendicular multiplied by a factor $1/\sin \Psi$. It may be noted that, for $\Psi = 90°$, the multiplicative factor is unity. Thus, the scattering function (3.119) can be cast in the following form:

$$T(\theta)_{EA} = x \int_0^{\pi/2} d\gamma \cos\gamma \cos(z \sin\gamma)[1 - \exp(\omega_{EA}^* \cos\gamma/\sin\Psi)], \quad (3.158)$$

for oblique incidence. Further, employing the extinction theorem (3.123), the extinction efficiency factor for a tilted cylinder can be expressed as:

$$Q_{ADA}^{ext} = \pi \operatorname{Re}[L(\omega^*/\sin\Psi)]. \quad (3.159)$$

These expressions have also been derived from an exact analytic solution of the problem (Sharma, 1989).

No numerical estimates of errors in angular-scattered intensity seem to have been made for a tilted cylinder. However, comparisons of exact and approximate extinction efficiency factors have been made by Stephens (1984) and by Fournier and Evans (1996). While the latter authors have included edge corrections in their calculations, the results of Stephens are based on the ADA without edge corrections. Both comparisons show that maxima and minima in the Q_{ADA}^{ext}-versus-x curve are out of phase with the exact curves for oblique incidence. Deviation increases with increasing obliqueness. The ADA extinction and absorption efficiencies for an absorbing cylinder have been examined by Cross and Latimer (1970) for normal as well as oblique incidence. The approximation was tested against exact theory for various sizes, refractive indexes and orientations with similar outcome.

It is assumed in writing (3.158) that the direction of propagation of rays in the scattering medium is the same as that of an incident ray. However, instead of this, if one assumes that after a ray strikes the boundary of the scatterer it is refracted and travels along this deviated path and not in the incident direction, then it can easily be seen that the phase shift function can be written as (Fournier and Evans, 1996):

$$\rho_{EADA}^* = 2x[(m^2 - \cos^2 \Psi)^{1/2} - \sin\Psi]. \quad (3.160)$$

The resulting approximation has been termed the "extended anomalous diffraction approximation" (EADA). Note that for $|m^2 - 1| \ll \sin^2 \Psi$, the right-hand side of (3.160) reduces to the corresponding expression for the standard ADA.

A simple empirical model that takes into account the edge behavior of small as well as large ρ_{EADA}^* for a given orientation has been demonstrated to be (Fournier and Evans, 1996):

$$Q_{edge} = \frac{0.996\,130}{(x^{2/3} + x_{crit})\sin^{2/3}\Psi}, \quad (3.161)$$

where

$$x_{crit} = \frac{3.6}{4|(m^2 - \cos^2 \Psi)^{1/2} - \sin \Psi|}. \tag{3.162}$$

The EADA improves the extinction efficiency dramatically even for large obliqueness. We shall return to a discussion of edge contributions again in Section 3.9.1.

The formulation of the EA or the ADA for an infinitely long circular cylinder can easily be expanded to include the scattering by an infinitely long cylinder of elliptic cross-section. Formally, the extinction efficiency factor for scattering by an infinitely long cylinder of elliptic cross-section is the same as that for an infinitely long circular cylinder (Fournier and Evans, 1996) and is given by (3.122). That is, the scattering efficiency factor can be written as:

$$Q_{ADA}^{ext} = \pi \, \text{Re}[H_1(\rho_{ADA}) + iJ_1(\rho_{ADA})], \tag{3.163}$$

where, from the geometry of the elliptic cross-section, it is found that:

$$\rho_{ADA} = \frac{2k(m-1)rb}{p \sin \Psi}, \tag{3.164}$$

and

$$p = (\cos^2 v + r^2 \sin^2 v)^{1/2}, \tag{3.165}$$

is the projection operator of the elliptic cross-section on the shadow plane, v is the angle between the semi-major axis of the ellipse and the plane defined by the direction of incident radiation and the infinite axis of the cylinder. The quantity r is the ratio of the semi-major axis a to the semi-minor axis b. A comparison of (3.163) with (3.122) shows that Q_{ADA}^{ext} for an elliptic cylinder is the same as that for a circular cylinder, but with x replaced by x/p.

The errors in the ADA for elliptic cylinders have not been contrasted with exact computations. However, exact calculations for forward and backward scattering were performed by Uzunoglu and Holt (1977). It was noted that the intensity functions have more maxima and minima in comparison with a circular scatterer of the same geometrical cross-section. It was suggested that this property could be used to determine the deviation from sphericity from measurement results.

3.6 COATED SPHERES

The coated sphere model has been extensively used in problems relating to the scattering of light by biological, atmospheric, other environmental and astrophysical particles. The exact solution of the problem has been given by Aden and Kerker (1951), Güttler (1952) and Shifrin (1952). The problem has been examined in the EA and the ADA by many workers (see, e.g., Morris and Jennings, 1977; Aas, 1984; Chen, 1987; Lopatin and Sid'ko, 1987; Zege and Kokhanovsky, 1989; Sharma and Somerford, 1991; among others). Let m_1, a_1 denote the refractive index and radius, respectively, of the core and m_2, a_2 denote the corresponding quantities of the coating. Then, the scattering function for the problem can be written as:

$$S(\theta)_{EA} = k^2 \left[\int_0^{a_1} b\, db\, J_0(qb) \left(1 - e^{ip_1\sqrt{a_1^2-b^2}+ip_2\left(\sqrt{a_2^2-b^2}-\sqrt{a_1^2-b^2}\right)}\right) \right.$$
$$\left. + \int_{a_1}^{a_2} b\, db\, J_0(qb) \left(1 - e^{ip_2\sqrt{a_2^2-b^2}}\right) \right], \qquad (3.166)$$

where $p_1 = k(m_1^2 - 1)$ and $p_2 = k(m_2^2 - 1)$. Note that when $p_1 = p_2$ equation (3.166) is nothing more than the scattering function for a homogeneous sphere of radius a_2 and refractive index m_2. Employing the change of variable, $b = a_1 \sin \gamma_1 = a_2 \sin \gamma_2$, the scattering function for a coated sphere may be cast in the following alternative form:

$$S(\theta)_{EA} = x_1^2 \int_0^{\pi/2} d\gamma_1 \frac{\sin 2\gamma_1}{2} J_0(qa_1 \sin \gamma_1) \left[1 - e^{i(p_1-p_2)a_1 \cos \gamma_1 + ip_2 a_2 \cos \gamma_2}\right]$$
$$+ x_2^2 \int_{\sin^{-1}(a_1/a_2)}^{\pi/2} d\gamma_2 \cos \gamma_2 \sin \gamma_2 J_0(qa_2 \sin \gamma_2) \left[1 - e^{ip_2 a_2 \cos \gamma_2}\right]. \quad (3.167)$$

Unlike the case of a homogeneous sphere where integration in the scattering function could be performed analytically to yield a closed form expression for forward scattering, integrations in (3.167) cannot be performed analytically even for forward scattering. However, simple expressions can be obtained for scattering and absorption efficiency factors and for angular scattering functions in special cases corresponding to thin and thick coatings (Morris and Jennings, 1977; Aragón and Elwenspoek, 1982; Bhandari, 1986). These special cases are of interest because they represent a number of realistic situations. For example, a biological cell can be treated as a thinly coated dielectric sphere. In this event, one is justified in assuming $a_1 \simeq a_2$ and $\cos \gamma_1 \simeq \cos \gamma_2$. Then, following Morris and Jennings (1977), one can express the scattering function as a sum of two terms. The first term is such that it can be treated exactly in the same manner as the scattering function for a homogeneous sphere. The second term is an analytic expression which, as expected, goes to 0 as $a_1/a_2 \to 1$. Similar observations have been made for a thickly coated sphere with the second term going to 0 as $a_1/a_2 \to 0$.

The extinction and absorption efficiency factors for a spherical particle with a core–mantel structure may be expressed as (see, e.g., Quirantes and Bernard, 2004):

$$Q_{ADA}^{ext} = 2 - 4\tilde{z}e^{-\tilde{z}\rho_1 \tan \beta_1} \frac{\cos \beta_1}{\rho_1} \sin(\tilde{z}\rho_1 - \beta_1)$$
$$- 4\left(\frac{\cos \beta_1}{\rho_1}\right)^2 e^{-\tilde{z}\rho_1 \tan \beta_1} \cos(\tilde{z}\rho_1 - 2\beta_1) + 4\left(\frac{\cos \beta_1}{\rho_1}\right)^2 \cos(2\beta_1)$$
$$- 4\frac{\cos \beta_2}{\rho_2} e^{-\rho_2 \tan \beta_2} \sin(\rho_2 - 2\beta_2) + 4\frac{\cos \beta_2}{\rho_2} \tilde{z}\, e^{-\tilde{z}\rho_2 \tan \beta_2} \sin(\tilde{z}\rho_2 - 2\beta_2)$$
$$- 4\left(\frac{\cos \beta_2}{\rho_2}\right)^2 e^{-\rho_2 \tan \beta_2} \sin(\rho_2 - 2\beta_2) + 4\left(\frac{\cos \beta_2}{\rho_2}\right)^2 \tilde{z}\, e^{-\tilde{z}\rho_2 \tan \beta_2} \sin(\tilde{z}\rho_2 - 2\beta_2),$$

$$(3.168)$$

and
$$Q_{ADA}^{abs} = 1 + \frac{\tilde{z}e^{-2\kappa_1 \tilde{z}}}{\kappa_1} + \frac{e^{-2\kappa_1 \tilde{z}} - 1}{2\kappa_1^2} + \frac{e^{-2\kappa_2} - \tilde{z}e^{-2\kappa_2}}{\kappa_2} + \frac{e^{-2\kappa_2} - e^{-2\kappa_2}}{2\kappa_2^2}, \qquad (3.169)$$
where
$$\tilde{z} = (1 - \xi^2); \quad \rho_1 = 2x_2(n_2 - 1),$$
$$\rho_2 = 2x_2[\xi n_1 + (1 - \xi)n_2 - 1],$$
$$\kappa_1 = 2x_2 n_2'; \quad \xi = \frac{a_1}{a_2},$$
$$\kappa_2 = 2x_2[\xi n_1' + (1 - \xi)n_2'].$$

The absorption is determined by:
$$\tan \beta_1 = \frac{n_2'}{n_2 - 1},$$
and
$$\tan \beta_2 = \frac{\xi n_1' + (1 - \xi)n_2'}{\xi n_1 + (1 - \xi)n_2 - 1}.$$

For $\xi = 1$, the above results reduce to that for a homogeneous sphere. Alternative general expressions for a two-layered sphere can also be found in Aas (1984).

Numerical comparisons of the EA and the ADA computations against the exact results have been performed for $i(\theta)$ and Q_{ext} for the scattering of unpolarized incident light. Chen (1987) made the comparisons for dielectric bubbles ($m_1 = 1, m_2 = m$) and wet dielectric balls ($m_2 = 1.33, m_1 = m$). Angular scattering in the EA was found to be in excellent agreement with exact results if $1 \leq |m| \leq 1.1$, $x \geq 50$ and $\theta \leq 10$. The errors are within 10% of exact results. However, the presence of Im$(n) \geq 0.04$ gives errors lower than 20% even if $|m|$ is as large as 2 and x as small as 5. The values $m_1 = 1.16$ and $m_2 = 1.02$ are typical of the refractive indexes of sand grains and biological material relative to water. Such grains occur in cohesive sediments in a variety of water bodies including estuaries. Sharma and Somerford (1991) examined the accuracy of the EA for Q_{ext} over the whole range of ξ values for such particles. The performances of the EA and the ADA were fairly accurate over the entire range of $x_2 \geq 10.0$. The maximum percent error in the EA and the ADA were 20% and 13%, respectively. For $x_2 > 500$, the percent error in both approximations was less than 6%.

The accuracy of the ADA for coated particles has also been examined by Zege and Kokhanovsky (1989). The domain examined is $1 \leq x_2 \leq 200$ for various values of m_1, m_2 and ξ. The following conclusions have been drawn (Kokhanovsky, 2004):

(i) The maximum error in the ADA is near the maximum of the extinction efficiency curve. The error could be reduced significantly by including edge corrections.
(ii) The accuracy of the ADA for extinction efficiency increases with increasing imaginary index of refraction of either core or coating and with decreasing ξ. A similar trend also holds for absorption efficiency.

(*iii*) The absorption efficiency is independent of n_1 and n_2 when the soft-particle condition is satisfied.
(*iv*) If $n'_1 = n'_2$ a soft-coated particle absorbs like a homogeneous particle of radius a_2.
(*v*) The absorption cross-section of a coated particle with a soft nonabsorbing coating is close to that of its core. This is because the coating itself does not absorb and produces little focusing or defocusing of incident radiation.
(*vi*) Absorption efficiency reaches a local maximum before falling to their large-size limit of ξ^2. This is in contrast to a monotonous increase in absorption efficiency from 0 to 1 for a homogeneous sphere.
(*vii*) The predictions of the ADA are better for a coated sphere than that for a homogeneous sphere. Recall that a similar observation was made for an infinitely long cylinder (at perpendicular incidence) as compared with a sphere.

A special case of coated sphere is the hollow sphere ($m_1 = 1$). The study of scattering and absorption by a hollow sphere is of particular interest in ocean optics. Aas (1984) has obtained closed form analytic expressions for scattering and absorption efficiencies for a hollow sphere in the framework of the ADA. The hollow sphere model in conjunction with the anomalous diffraction approximation has also been used for studying nematic droplets (Huang et al., 1996). The extinction efficiency for a hollow sphere is:

$$Q^{ext}_{ADA} \simeq 2\,\mathrm{Re}\left[\frac{4}{\rho_2^2}(-ise^{is} + e^{is} - 1) + \frac{1}{\rho_2^2}(-iSe^{iS} + e^{iS} + ise^{is} - e^{is})\right.$$
$$\left. + \frac{s^4}{2\rho_2^2}\left(\frac{e^{iS}}{S^2} + i\frac{e^{iS}}{S} - \frac{e^{is}}{s^2} - \frac{ie^{is}}{s}\right) + \frac{s^4}{2\rho_2^2}(E_1(is) - E_1(iS))\right], \quad (3.170)$$

where

$$s = 2x_2(n-1)(1-\xi^2)^{1/2}, \quad S = 2x(x_2(n-1)(1-\xi)),$$

and

$$E_1(is) = -Ci(s) + i\left(Si(s) - \frac{\pi}{2}\right), \quad (3.171)$$

with

$$Ci(s) = -\int_s^\infty \frac{\cos t}{t}\,dt, \quad (3.172)$$

as the Cosine integral and:

$$Si(s) = \int_0^s \frac{\sin t}{t}\,dt, \quad (3.173)$$

as the Sine integral. For absorption efficiency one obtains:

$$Q_{abs} \simeq 1 + \frac{1}{2\kappa_2^2}(t\,e^{-t} + e^{-t} - 1) + \frac{1}{\kappa_2^2}(T\,e^{-T} + e^{-T} - t\,e^{-t} - e^{-t})$$
$$+ \frac{t^4}{16\kappa_2^2}\left(\frac{e^{-T}}{T} - \frac{e^{-T}}{T^2} - \frac{e^{-t}}{t} + \frac{e^{-t}}{t^2}\right) + \frac{t^4}{16\kappa_2^2}(E_1(t) - E_1(T)), \quad (3.174)$$

where
$$t = 2\kappa_2(1 - \xi^2)^{1/2}$$
and
$$T = 2\kappa_2(1 - \xi).$$

The function E_1 is:
$$E_1(t) = \int_t^\infty \frac{e^{-x}}{x}\,dx. \qquad (3.175)$$

For $\xi = 0$, $s = S = \rho_2$ and $t = T = 2\gamma$, and it can be easily verified that (3.170) and (3.174) reduce to the corresponding expressions for a homogeneous sphere.

The solution of the scattering problem for a multi-layered sphere in the framework of the ADA is a straightforward extension of the procedure followed for a coated particle. The condition of softness now needs to be satisfied in each layer. Exact solutions for a sphere with more than one coating have been given by Mikulski and Murphy (1963) and Wait (1963).

3.7 SPHEROIDS AND ELLIPSOIDS

The problem has been examined analytically as well as numerically by Chen (1995) and later by Chen and Yang (1996). A notable outcome from these studies is the realization that – in the framework of the EA – the scattering function for a dielectric spheroid $S_{EA}^{oid}(a, b, m)$ can be related analytically to the scattering function by a sphere $S_{EA}^{ere}(a_{eff}, m_{eff}, \theta)$ in the following way:

$$S_{EA}^{oid}(a, r, m, \theta) = \frac{\alpha_1}{\beta_1^2} S_{EA}^{ere}(a_{eff}, m_{eff}, \theta), \qquad (3.176)$$

where a and b are semi-major and semi-minor axes, $r = a/b$ is the aspect ratio of the spheroid and m is its refractive index. The radius and refractive index of the equivalent sphere denoted, respectively, by a_{eff} and m_{eff} are given by the relations:

$$a_{eff} = \frac{\beta_1}{r} a, \qquad (3.177)$$

and

$$m_{eff} = 1 + \frac{r}{\alpha_1 \beta_1}(m - 1). \qquad (3.178)$$

If the semi-major axis is defined by coordinates (a, θ_0, ϕ_0), α_1 and β_1 are given by the relations:

$$\beta_1 = \sqrt{\alpha_1^2 R^2 + S^2}, \qquad (3.179)$$

and

$$\alpha_1 = \sqrt{U^2 + r^2 V^2}, \qquad (3.180)$$

where $U^2 + V^2 = R^2 + S^2 = 1$ and U, V, R and S are related to θ_0, ϕ_0 and the scattering angle θ as:

$$R = \frac{\cos(\theta/2)\cos\theta_0\cos\phi_0 - \sin(\theta/2)\cos\theta_0}{V}, \quad (3.181)$$

and

$$U = \cos(\theta/2)\cos\theta_0 + \sin(\theta/2)\sin\theta_0\cos\phi_0. \quad (3.182)$$

The resulting scattering function has been examined analytically as well as numerically by Chen (1995) and Chen and Yang (1996). A notable feature of these studies is that – although the relation (3.176) has been derived within the framework of the EA – numerical tests of angular intensity as well as extinction efficiencies for a dielectric spheroid from the relation (3.176) have shown it to be valid for exact T-matrix solutions too. In other words, this means that the scattering function for a spheroid can be computed employing Mie theory for a homogeneous dielectric sphere of radius a_{eff} and refractive index m_{eff}. The method, therefore, provides a fast and easy way to forecast the scattering quantities of a spheroid. The results are particularly good when incident light is perpendicular to the symmetry axis. The domains of validity of this equivalence are found to be $kb^2/a \geq 4$ for a prolate spheroid and $ka^2/b \geq 4$ for an oblate spheroid when $\theta < 30°$. The origin of these inequalities has been shown to lie in the restriction on the radius of curvature of the spheroid.

The reliability of (3.176) has been evaluated against the exact T-matrix computations for the scattering of unpolarized light by a nonabsorptive prolate water droplet of $ka = 20$, and $a/b = 1.25$ obtained employing the framework of modified generalized EA for a large particle. The following conclusions have been drawn regarding the domain of validity of this approximate formula.

(*i*) The MGEA results agree well with T-matrix calculations for scattering angles up to about 10°. The agreement continues to be fairly good for scattering angles up to about 30°.
(*ii*) For a prolate spheroid the MGEA works better when the direction of incident light is normal to the major axis. On the other hand, for an oblate spheroid, the MGEA formula works better when the direction of incident light is parallel to the major axis. This is because light enters the spheroid mostly through smooth regions of surface. A similar observation regarding direction of incidence has been reported by Holt and Shepherd (1979) when inquiring into the validity of the ADA for scatterings at 0° and 180°.
(*iii*) The restrictions on curvature – namely, $kb^2/a \geq 4$ for a prolate (or $ka^2/b \geq 4$ for an oblate spheroid), in conjunction with (3.178) – imply that the formula is valid for a prolate spheroid if $r < 1/(m-1)$ (and for an oblate spheroid if $r > m - 1$).

The EADA for scattering by an infinite long cylinder was described in Section 3.5.6. It differs from the standard ADA in that it takes into account refraction at the boundary of the particle. It can easily be verified that the EADA for a prolate spheroid consists in replacing the usual ρ^*_{ADA} by:

$$\rho^*_{EADA} = 2kb\left[(m^2 - \cos^2\Psi)^{1/2} - \sin\Psi\right], \tag{3.183}$$

and for an oblate spheroid it consists in replacing ρ^*_{EA} by:

$$\rho^*_{EADA} = 2ka\left[(m^2 - \cos^2\Psi)^{1/2} - \sin\Psi\right]. \tag{3.184}$$

Fournier and Evans (1991) have investigated the efficacy of the EADA in predicting extinction efficiency factors for an oriented spheroid. In this paper Fournier and Evans refer to this approximation as the "extended eikonal approximation", but in later papers the same approximation is referred to as the "anomalous diffraction approximation". Numerical studies show that for prolate spheroids with $r > 3$ the results from extended eikonal approximation are least accurate when the orientation angle Ψ is close to $0°$. On the other hand, for oblate spheroids with $r < 1/3$, the maximum error occurs near $\Psi = \pi/2$.

The same problem has also been addressed by Chen et al. (2003, 2004) for a prolate spheroid. The study was performed for three electromagnetic wave modes. (i) TM mode, where the magnetic field vector is perpendicular to the major axis of the spheroid and the electric field vector is parallel to the major axis. (ii) TE mode, where the electric vector is perpendicular to the major axis of the spheroid and the magnetic vector is parallel to it. (iii) TEM mode, where electric as well as magnetic vectors are perpendicular to the major axis of the spheroid. The extinction efficiency for TE and TM mode may be stated as:

$$Q^{ext}_{ADA} = \frac{4}{\pi ab}\left[(\pi ab/2)\{1 - 2m\sin\rho^*_b/\rho^*_b + 4m\sin^2(\rho^*_b/2)/\rho^{*}_b\} + (\pi ab/2)\{k(ab^2)^{1/3}/2\}^{-2/3}\right], \tag{3.185}$$

where $\rho^*_b = ka(m-1)$. The first term in square brackets on the right-hand side is the usual ADA term. The second term on the right-hand side is the contribution of the edge term taken from the work of Nussenzveig and Wiscombe (1980). For TEM mode, the extinction efficiency can be written as:

$$Q^{ext}_{ADA} = \frac{4}{\pi b^2}\left[(\pi b^2/2)\{1 - 2m\sin\rho^*_a/\rho^*_a + 4m\sin^2(\rho^*_a/2)/\rho^{*}_a\} + (\pi ab/2)\{k(ab^2)^{1/3}/2\}^{-2/3}\right]. \tag{3.186}$$

In the above equation $\rho^*_a = ka(m-1)$. Expression (3.185) is expected to be good if:

$$kb(m-1)\delta L(\mathbf{r}) = \rho^*_b \delta L(\mathbf{r}) < \pi/2, \tag{3.187}$$

where $L(\mathbf{r})$ is the path of the radiation inside the scatterer and $\delta L(\mathbf{r})$ accounts for the deviation of $L(\mathbf{r})$ from the path in the corresponding equiphase sphere. Similarly, the validity of this approximation for TEM mode requires:

$$ka(m-1)\delta L(\mathbf{r}) = \rho^*_a \delta L(\mathbf{r}) < \pi/2. \tag{3.188}$$

After some algebraic manipulation, these conditions can be recast in a more tractable form. For TE and TM modes one gets:

$$\frac{16b}{\pi^2 \lambda} \frac{m^2-1}{m} \left(1 - \frac{b}{a}\right)\left(1 + \frac{a^2}{b^2}\right)^{-1} < 1, \qquad (3.189)$$

and for TEM mode this is given by:

$$\frac{16a}{\pi^2 \lambda} \frac{m^2-1}{m} \left(1 - \frac{a}{b}\right)\left(1 + \frac{b^2}{a^2}\right)^{-1} < 1. \qquad (3.190)$$

These restrictions are in addition to the usual conditions required for the validity of the ADA. It is clear from (3.189) and (3.190) that the validity of approximation improves as the spheroid's curvature decreases – that is, as $a \to b$ – and also as the relative refractive index approaches unity.

The periodicity of oscillatory behavior in (3.185) and (3.186) is the same as that for extinction efficiency in the ADA for a homogeneous sphere. Thus, a homogeneous sphere with phases given by ρ_b^* and ρ_a^* will obviously reproduce the correct oscillatory behavior as the extinction efficiency of the spheroid. This observation, indeed, forms the basis of an approximation known as the "equiphase approximation". It has been suggested that this approximation can be used for a particle of arbitrary shape. The prescription is to write the extinction efficiency for a nonspherical particle as (Chen *et al.*, 2004) the sum of two terms:

$$Q_{EPA}^{ext} = Q_{ext}(v) + Q(s), \qquad (3.191)$$

where $Q_{ext}(v)$ represents the contribution of an equivalent sphere defined as a sphere for which the phase shift suffered by the central ray is equal to the maximum phase shift of light passing through the non-spherical particle. The term $Q(s)$ represents edge correction. Numerical comparisons of the extinction efficiency for a nonabsorbing spheroid, calculated on the basis of (3.191), show that this approximation is in good agreement with rigorous numerical computations.

The extinction and absorption efficiency factors of an ellipsoidal particle as predicted in the framework of the ADA have been examined, among others, by Lind and Greenberg (1966), Latimer (1975, 1980), Paramonov *et al.* (1986), Lopatin and Sid'ko (1988), Paramonov (1994), Streekstra (1994) and Belafhal *et al.* (2002). Here, too, it is found that the formulas for a spherical particle can be used to obtain the corresponding quantities for an ellipsoid. One only needs to change the diameter of the spherical particle to the value (Kokhanovsky, 2005):

$$h = \frac{2abc}{\sqrt{(bc \cos \mu_1)^2 + (ac \cos \mu_2)^2 + (ab \cos \mu_3)^2}} = \frac{3V}{2P}, \qquad (3.192)$$

where a, b, c are the semi-axes of the ellipsoidal particle, and μ_1, μ_2, μ_3 are angles which the incident radiation makes with the coordinate axes x, y and z respectively. The symbol V represents the volume of the particle, P is its geometrical cross-section and h is the maximal length of a ray inside the particle. This approximation is, therefore, nothing more than a kind of equiphase approximation. The problem of scattering of light by absorbing and non-absorbing spheroidal particles has been treated in the WKBA too (Belafhal et al., 2002).

The angular scattering pattern of light scattered by an arbitrarily oriented ellipsoidal particle has been investigated in great detail by Streekstra (1994) in the context of scattering of light by red blood cells. A comparison of the ADA with T-matrix calculations shows that the ADA is highly accurate for ellipsoidal red blood cells within the scattering angle of about 15°.

3.8 SOME OTHER SHAPES

Following scattering function (3.20) and extinction theorem (3.7), the extinction efficiency factor for an arbitrarily shaped particle in the ADA formulation can be formally expressed as:

$$Q_{ADA}^{ext} = \frac{2}{P} \text{Re} \left[\iint_P [1 - e^{ikl(m-1)}] \, dP \right]. \tag{3.193}$$

Similarly, following (3.50), the absorption efficiency factor can be formally expressed as:

$$Q_{ADA}^{abs} = \frac{1}{P} \iint_P [1 - e^{-kln'}] \, dP. \tag{3.194}$$

In the above equations, P is the projected area of the particle on a plane perpendicular to the incident direction and l is the geometrical path length of radiation traveling undeviated inside the particle. In the following subsections we give explicit expressions for some of the commonly used shapes.

3.8.1 Columnar particles

Consider a hexagonal column whose axis of symmetry is normal to the direction of incident radiation ($\Psi = 0$). Then, for edge-on incidence the extinction efficiency can be obtained unambiguously (Chýlek and Klett, 1991a, b) as:

$$Q_{ADA}^{ext} = 2 + \frac{4e^{-\kappa}(\kappa \cos \rho - \rho \sin \rho)}{\rho^2 + \kappa^2} - \frac{2e^{-\kappa/2}}{\rho^2 + \kappa^2} \kappa \cos(\rho/2) \sin(\rho/2), \tag{3.195}$$

and the absorption efficiency is simply:

$$Q_{ADA}^{abs} = 1 - e^{-\kappa} + \frac{e^{-2\kappa}}{\kappa}, \tag{3.196}$$

where $\rho = 2ka(n-1)$ and $\kappa = 2kan'$ with a as one side of the hexagon. When incidence is on a flat orientation, equations (3.193) and (3.194) yield extinction and absorption efficiency factors given by:

$$Q_{ADA}^{ext} = 2 - \frac{\kappa'}{\rho'^2 + \kappa'^2} - e^{-\kappa'}\cos(\rho') + e^{\kappa'}\frac{(\kappa'\cos(\rho') - \rho'\sin(\rho'))}{\rho'^2 + \kappa'^2}, \quad (3.197)$$

and

$$Q_{ADA}^{abs} = 1 - \frac{e^{-2\kappa'}}{2} - \frac{1 - e^{-2\kappa'}}{4\kappa'}, \quad (3.198)$$

where $\kappa' = (\sqrt{3}/2)\kappa$ and $\rho' = (\sqrt{3}/2)\rho$. The expressions for the extinction and absorption efficiencies of radiation by finite hexagonal cylinders have been obtained by Kuznetsov and Pavlova (1988). Sun and Fu (1999) have derived analytic formulas for absorption and extinction efficiency factors for a hexagonal column which is in arbitrary orientation. The formulas take a rather cumbersome form and for this reason are given separately as an appendix (Appendix A).

Analytic equations for the scattering and absorption properties of a column consisting of an arbitrary polygonal base have been obtained by Chýlek and Klett (1991a). Integrations appearing in formal expressions (3.193) and (3.194) can be evaluated by dividing the projected area (over which integration is to be performed) into an arbitrary number of nonoverlapping projected areas. To see the advantage achieved in approaching the problem in this way, consider two nonoverlapping sections. The integrand to be considered is of the form:

$$1 - e^{\mathcal{P}_1 + \mathcal{P}_2} = (1 - e^{\mathcal{P}_1}) + (1 - e^{\mathcal{P}_2}) - (1 - e^{\mathcal{P}_1})(1 - e^{\mathcal{P}_2}).$$

As the sections of the particle are nonoverlapping the last term on the right-hand side vanishes. Hence, the total extinction efficiency factor can be expressed simply as a sum of the two extinction efficiency factors. The argument can be extended to N nonoverlapping sections. As a result the extinction and absorption cross-sections of the whole particle can be written as the following sum:

$$C_{ADA}^{ext} = \sum_{i=1}^{N} C_{ADA}^{ext}(i), \quad C_{ADA}^{abs} = \sum_{i=1}^{N} C_{ADA}^{abs}(i), \quad (3.199)$$

where $i = 1$ to N are N nonoverlapping sections of a particle. The result has been referred to as the *addition theorem*.

Chýlek and Klett (1991a, b), employing the addition theorem, have studied the scattering of light by columnar particles with triangular, trapezoidal, hexagonal or polygonal bases and obtained analytic expressions for Q_{ADA}^{ext} and Q_{ADA}^{abs}. Study of the effect of type of base on the scattering of normally incident radiation by a columnar particle allows one to make the following observations:

(i) The maximum absorption is for a parallelogram column. The minimum absorption is for a triangular column.
(ii) For single-scattering albedo, the parallelogram column shows the maximum oscillations and the triangular column shows the minimum oscillations for lower absorption ($m = 1.1 + i0.01$). For higher absorption ($m = 1.1 + i0.1$), the oscillations disappear.

(iii) All absorption efficiencies show a tendency to approach their large-particle asymptotic limit, which is 1.
(iv) The absorption and extinction efficiencies in the ADA are found to be within 25% of the exact values for refractive indexes with a real part less than 1.40 and within 10% for refractive indexes with a real part less than 1.20.

Exact solutions for the scattering of electromagnetic waves by a three-dimensional hexagonal column have been obtained by Fu et al. (1999), using the finite difference time domain technique. On the other hand, Mano (2000) has used the boundary element method to obtain exact results. Baran and Havemann (2000) have calculated the absorption efficiency and single-scattering albedo of hexagonal columns, using a generalization of separation of variable method for size parameter of up to 80.

3.8.2 Cube

Another particle shape for which the extinction and absorption efficiency predictions of the ADA have been examined in detail is a cube (Flatau, 1992; Maslowska et al., 1994). Two orientations of the cube have been investigated: (i) side incidence and (ii) edge incidence. In the first case incidence is perpendicular to the face of the cube and in the latter case incidence is perpendicular to the edge of the cube (a rotation of 45° from side incidence). The computations from approximate formulas have been compared with discrete dipole approximation results. The discrete dipole approximation was initiated by Purcell and Pennypacker (1973), but was considerably extended mainly by Draine (1988), Flatau (1992) and Draine and Flatau (1994). For the extinction efficiency factor, comparisons for side incidence show fairly good agreement between the exact and the ADA methods. The errors are found to be of the same order as that observed between Mie results and the ADA predictions for a sphere. In contrast, for edge-on incidence, deviations from exact results are much larger for the same refractive index. The maximum error is about 80% for $x = 11.5$ and $m = 1.33 + i0.01$. It is noted that the first maximum in the extinction curve, predicted at $x = 6$ by the ADA, actually occurs at $x = 7$. For a less absorbing cube the disagreement becomes more pronounced. The absorption efficiency is well-approximated in the ADA for face-on as well as for side-on incidence. Errors are similar to those in the ADA for a homogeneous sphere. It has been suggested that additional empirical corrections similar to edge corrections should result in better accuracy. However, further work does not seem to have been done in this direction.

3.8.3 Plates and needles

Plates and needles are useful models of some atmospheric and oceanic particles. For example, mica appears as thin flakes and amphiboles may have needle-like shapes. Several phytoplankton species have disk-like shapes. Ice crystals may occur in the form of needles and plates. The problem of scattering and absorption of radiation by such scatterers can be considered as special cases of electromagnetic scattering by a three-dimensional hexagonal or circular column. For a hexagonal plate of thickness t

and side a in a horizontal orientation (i.e., the hexagonal side normal to the direction of incident radiation), the absorption and extinction efficiency factors can be written as (Chýlek and Klett, 1991b; Chýlek and Videen, 1994):

$$Q_{ADA}^{ext} = 2[1 - \cos[kt(n-1)]\,e^{-ktn'}], \qquad (3.200)$$

and

$$Q_{ADA}^{abs} = [1 - e^{-2ktn'}]. \qquad (3.201)$$

The above expressions also describe the extinction and absorption efficiency factors for a circular plate. The cross-sections, however, are different in the two cases because of different geometrical cross-sections, whereas the projected area of a hexagonal plate is $3\sqrt{3}a^2/2$ – for a circular plate this factor is πa^2. Results from the above expressions (circular plate) have been compared with exact (DDA) results by Chýlek and Videen (1994). In these comparisons $t = 0.35$ mm and $a = 2$ mm. The three refractive indexes chosen were $m = 1.231 + i0.000\,7531$, $m = 1.251 + i0.000\,9563$ and $m = 1.273 + i0.001\,1280$. The error in absorption efficiency factor was noted to be around 5%, while the error in extinction and scattering efficiency factor was found to be less than 12%. The error in albedo was found to be less than 2%.

3.8.4 Parallelepiped

A method for predicting the light scattering and absorption by non-absorbing soft particles of arbitrary shape and size has recently been put forward by Rysakov (2004, 2006). The procedure regards the interaction of light with a particle as consisting of two processes. The first process considers scattering in its narrow sense. That is, it considers scattering as the process of secondary radiation as in the case of the Rayleigh approximation or the RGDA. Rysakov has extended this concept to the realm of the WKBA method. In this generalization one can write the scattered field as:

$$E_s(\theta, \varphi) = \frac{3(m^2 - 1)}{4\pi(m^2 + 1)} \frac{\chi}{\chi_1} \rho^2 f(\rho', \theta', \varphi') \sqrt{2Z_0 P_0} \sqrt{1 - \sin^2\theta \cos^2\varphi}, \qquad (3.202)$$

where

$$f(\rho', \theta', \varphi') = \frac{1}{V} \int m(r)\, e^{i\rho' r} r^2 \sin\theta\, \partial\theta\, \partial\varphi\, \partial r, \qquad (3.203)$$

with $\chi = V/a^3$ (a being the characteristic dimension of the particle) as the normalized volume and χ_1 as the projected area. Further:

$$\theta' = \arctan\left[\frac{\sqrt{\cos^2\phi_B \sin^2\phi_B + (\cos\phi_B \sin\tilde{\theta} - \cos\tilde{\theta}\sin\theta_B \sin\phi_B)^2}}{\cos\phi_B \cos\tilde{\theta} + \sin\tilde{\theta}\sin\theta_B \sin\phi_B}\right], \qquad (3.204)$$

$$\phi' = -\arctan\left[\cot\theta_B \sec\phi_B \sin\tilde{\theta} - \cos\tilde{\theta}\tan\phi_B\right], \qquad (3.205)$$

$$\rho_B = 2\rho\sqrt{m\sin^2(\theta/2) + \frac{(m-1)^2}{4}}, \qquad (3.206)$$

$$\theta_B = \arcsin\left(\frac{\theta}{2\rho_B}\right), \qquad (3.207)$$

and
$$\phi_B = \varphi, \tag{3.208}$$

with $\tilde{\theta}$ as a turn on a plane normal to the direction of the incident wave. The second process to contribute is the diffraction at the edges of the particle. With length $2a\Gamma$ along the z-axis, one has:

$$f(\rho', \theta', \varphi') = \frac{\sin(\rho'\cos\theta'\sin\phi')}{\rho'\sin\theta'\sin\phi'} \frac{\sin(\rho'\cos\theta'\cos\phi')}{\rho'\sin\theta'\cos\phi'} \frac{\Gamma\rho'\cos\theta'}{\Gamma\rho'\cos\theta'}, \tag{3.209}$$

$$\chi = 8\Gamma, \tag{3.210}$$

$$\chi_1 = 4(\cos\tilde{\theta} + \Gamma\sin\tilde{\theta}). \tag{3.211}$$

Relevant expressions for a sphere and a cylinder have also been given by Rysakov (2006). Numerical calculations show that the error for $m \approx 1.33$ is approximately the same ($\approx 20\%$) as with the standard ADA. However, a significant advantage of this technique is that the approach in question allows calculation of not only the integral characteristics but also of the angular scattering pattern.

3.8.5 Statistical interpretation of the ADA

It is evident from a look at (3.193) and (3.194) that the numerical outcome of these equations will be independent of the order in which contributions from the ray path in the integral are computed. Thus, by dividing the projected area into equal area elements and counting the geometrical paths according to their lengths, a probability function $p(l) \, dl$ can be found that gives the probability of length l between l and $l + dl$. This interpretation of the ADA allows (3.193) and (3.194) to be re-phrased as (Xu, 2003; Xu et al., 2003):

$$Q_{ADA}^{ext} = 2\,\mathrm{Re}\left[\int [1 - e^{ikl(m-1)}] p(l) \, dl\right], \tag{3.212}$$

and

$$Q_{ADA}^{abs} = \int [1 - e^{-kln'}] p(l) \, dl, \tag{3.213}$$

with the normalization $\int p(l) \, dl = 1$. The percentage of particle area that corresponds to the specific geometric path in the above equations is independent of the particle's physical size if the shape and aspect ratio of the particle remain the same. If we denote by $p_0(l)$ the geometrical path distribution of rays for one particle with unit size, the ray distribution for a particle of the same shape and orientation, but a different size L, can be obtained by scaling the old l by l/L. This leads to the relation $dl \to dl/L$. Hence, the probability function $p_0(l) \, dl$ for the new scatterer becomes:

$$p(l) \, dl = \frac{1}{L} p_0(l/L) \, dl.$$

The geometrical path distribution has been examined by Xu et al. (2003) and Xu (2003) for a spheroid, a system of randomly oriented monodisperse spheroids, a system of polydisperse spheroids in a fixed orientation and polydisperse spheroids in a random orientation. For an isolated spheroid it has been shown that:

$$p_0(l) = \frac{1}{2r^2b^2}(\sin^2\Psi + r^2\cos^2\Psi)lH\left[\frac{2rb}{(\sin^2\Psi + r^2\cos^2\Psi)^{1/2}} - l\right] \quad l \geq 0, \quad (3.214)$$

where $H(x)$ is the Heaviside step function, which is 0 for $x < 0$ and 1 for $x > 0$. Further, the ray distribution for a system of spheroids at a fixed orientation Ψ with a log-normal size distribution:

$$f(a) = \frac{1}{(2\pi)^{1/2}\sigma}a^{-1}\exp\left[\frac{-\ln^2(a/a_m)}{2\sigma^2}\right], \quad (3.215)$$

can be calculated to be (Xu et al., 2003 and Xu, 2003):

$$p_{poly}(l) = \frac{(r^{-2}\sin^2\Psi + \cos^2\Psi)l}{4}\frac{\text{erfc}[(1/\sqrt{2}\sigma)\ln[(r^{-2}\sin^2\Psi + \cos^2\Psi)^{1/2}l/(2a_m)]]}{a_m^2\exp(2\sigma^2)}, \quad (3.216)$$

where erfc(x) is the complementary error function and $\int_0^\infty f(a)\,da = 1$. The subscript *poly* denotes polydispersion. The random orientations and polydispersion both tend to wash out the characteristic features of an individual particle and the probability density function then appears to be approaching Gaussian in nature. Similar observations have been made for a finite cylinder. The shape characteristics further spread out if particles of different shapes are involved. Thus, for randomly oriented polydisperse particles one may use a Gaussian probability distribution function for the geometrical paths:

$$p(l) = \frac{1}{\sqrt{2\pi}\sigma}\exp\left[-\frac{(l-\mu)^2}{2\sigma^2}\right], \quad (3.217)$$

in (3.212) and (3.213). Integration in these expressions can be performed to yield:

$$Q_{GRA}^{ext} = 2 - 2\cos[k(n-1)(\mu - k\sigma^2 n')]\exp\left[-k\mu n' - \frac{k^2\sigma^2[(n-1)^2 - n'^2]}{2}\right], \quad (3.218)$$

and

$$Q_{GRA}^{abs} = 1 - \exp[-2kn(\mu - kn'\sigma^2)]. \quad (3.219)$$

The approximation has been termed as the Gaussian ray approximation (GRA) (Xu et al., 2003).

For intermediate size particles – that is, if $kl(n-1) \ll 1$ and $kln' \ll 1$ – equations (3.218) and (3.219) reduce to the following expressions:

$$Q_{ext} = 2kn'\langle l\rangle + k^2[(n-1)^2 - n'^2]\langle l^2\rangle, \quad (3.220)$$

and

$$Q_{abs} = 2kn'\langle l\rangle - 2k^2 n'^2\langle l^2\rangle, \quad (3.221)$$

where $\langle l\rangle = \mu$ and $\langle l^2\rangle = \mu^2 + \sigma^2$. Equations (3.220) and (3.221) are in agreement with corresponding expressions obtained in the RGDA. The expressions are also in agreement with those obtained by Chýlek and Li (1995) from the ADA. In this limit, the RGDA and the ADA yield identical expressions for extinction and absorption efficiency factors. Numerical comparisons of (3.218) and (3.219) with Mie results for a sphere of $m = 1.05 + i0.0005$ have also been performed. It is found that the absorption efficiency calculated engaging GRA concurs very well with Mie results. It differs at most by 2% from the exact calculations. The extinction efficiencies in the geometric ray approximation, however, agree well with the exact results only for intermediate size spheres. The errors become quite large at larger sizes. For polydisperse particles, the results of GRA are better. The maximum relative deviation compared with the ADA is 3.5%. The maximum relative error compared with the Mie theory is 7%.

By defining a dimensionless quantity $\tilde{l} = l/l_{max}$, where l_{max} is the maximum ray length for a particle with a given orientation, the extinction and absorption efficiency factors given by (3.212) and (3.213) can be re-cast as follows (Yang et al., 2004):

$$Q_{ADA}^{ext} = 2\,\mathrm{Re}\left[\int_0^1 \left[1 - e^{ikl_{max}\tilde{l}(m-1)}\right]p(\tilde{l})\,d\tilde{l}\right], \quad (3.222)$$

and

$$Q_{ADA}^{abs} = \int_0^1 \left[1 - e^{-kl_{max}\tilde{l}n'}\right]p(\tilde{l})\,d\tilde{l}. \quad (3.223)$$

In the above equations it is assumed that $\tilde{l}_{min} = 0$. If this is not so, the lower limit of the integral could be replaced as l_{min}/l_{max}. Further, it is convenient to define a cumulative fraction associated with $p(\tilde{l})$ as:

$$q(\tilde{l}) = \int_0^{\tilde{l}} p(\tilde{l})\,d\tilde{l}. \quad (3.224)$$

Since $p(\tilde{l}) = 0$ at $\tilde{l} = 0$, the differentiation of $q(\tilde{l})$ with respect to $\tilde{l}$ yields:

$$\frac{dq}{d\tilde{l}} = p(\tilde{l}) \quad \text{or} \quad dq = p(\tilde{l})\,d\tilde{l}. \quad (3.225)$$

Since $p(\tilde{l}) \geq 0$ for all $\tilde{l}$, $q(\tilde{l})$ defined by (3.224) is a monotonically increasing function of $\tilde{l}$ and, hence, there exists one-to-one mapping between $\tilde{l}$ and q. Consequently, $\tilde{l}$ can be expressed as a function of q. In other words, one can write $\tilde{l}_q = \tilde{l}(q)$. This allows one to re-define (3.222) and (3.223) as (Yang et al., 2004):

$$Q_{ADA}^{ext} = 2\,\mathrm{Re}\left[\int_0^1 \left[1 - e^{ikl_{max}\tilde{l}_q(m-1)}\right]dq\right], \quad (3.226)$$

and

$$Q_{ADA}^{abs} = \int_0^1 \left[1 - e^{-kl_{max}\tilde{l}_q n'}\right]dq. \quad (3.227)$$

The expressions (3.222), (3.223), (3.226) and (3.227) are independent of particle size. This implies that – in problems of polydisperse particles – the computational resources required should become considerably fewer.

For a homogeneous sphere, it can easily be demonstrated that $\tilde{l}_q = q^{1/2}$ (Yang et al., 2004). This allows one to develop the extinction and absorption efficiencies for a sphere as:

$$Q^{ext}_{ADA} = 2\,\text{Re}\left[\int_0^1 \left[1 - e^{ikl_{max}q^{1/2}(m-1)}\right]dq\right], \quad (3.228)$$

and

$$Q^{abs}_{ADA} = \int_0^1 \left[1 - e^{-kl_{max}q^{1/2}n'}\right]dq. \quad (3.229)$$

Identifying $q = \sin^2\gamma$ and $kl_{max}(m-1) = \rho^*$, it may easily be seen that (3.228) and (3.229) comply with the standard anomalous diffraction formula.

The effectiveness of this approach has been examined by Yang et al. (2004) who contrast numerically the results obtained by this approach and by the anomalous diffraction approximation for a side-on incidence hexagonal column. The agreement is remarkable for both absorption and extinction efficiencies.

Among other irregular shape particles for which light scattering has been treated in the ADA or the EA are: (*i*) homogeneous spheres with radial projections (Latimer, 1984a) and homogeneous spheres with holes (Latimer, 1984b); (*ii*) large scattering objects whose axis of symmetry is not parallel to the direction of incident wave vector (Bourrely et al., 1989); (*iii*) particles with a rough surface (Bourrley et al., 1986a, b).

3.9 RANDOMLY ORIENTED MONODISPERSE PARTICLES

In a large number of practical applications, interest lies in the scattering and absorption characteristics of a dispersion of particles in random orientations. In this section we consider this aspect of the scattering problem and examine the role played by the ADA and the EA in the solution of some commonly occurring nonspherical particle shapes. We limit ourselves to a scattering volume in which there is no multiple scattering and, hence, each scatterer can be treated independently. Consequently, the averaged quantities can be obtained by integrating the quantity in question over all possible directions of the symmetry axis. This constitutes part of the solution to a larger radiative transfer problem which takes into account the effect of multiple scattering.

3.9.1 Long-circular and elliptic cylinders

For random orientations, angle averaging for circular cylinders may be defined as (see, e.g., Fournier and Evans, 1996; Aas, 1984):

$$\bar{Q}_{ext} = \frac{\int_0^{\pi/2} Q_{ext}\sin^2\Psi\,d\Psi}{\int_0^{\pi/2}\sin^2\Psi\,d\Psi}, \quad (3.230)$$

where Ψ is the angle between the z-axis of the cylinder and the incident wave direction. Orientation averaging for absorption efficiency can also be defined in a similar manner. For infinitely long cylinders, Q^{ext}_{ADA} has an analytic form and the replacement of Q_{ext} in (3.230) by Q^{ext}_{ADA} allows integrations in (3.230) to be performed analytically. The final form given by Fournier and Evans (1996) is:

$$\bar{Q}^{ext}_{ADA} \sim 2 + e^{-2\kappa x}\left[\frac{4}{3}\rho^2\left(1 - \frac{\rho\pi}{4}\left[J_0^2(\rho/2) - \frac{2}{\rho}J_0(\rho/2)J_1(\rho/2) + \left(1 - \frac{2}{\rho}\right)J_1(\rho/2)\right]\right) - 2\right]. \tag{3.231}$$

The equation is readily computable. For large ρ, the asymptotic expansion of Bessel functions allows (3.231) to be represented as:

$$\bar{Q}^{ext}_{ADA} \sim 2 - \frac{4\cos\rho}{\rho} + \frac{3}{2}\left(\frac{1 - 4\sin\rho}{\rho^2}\right). \tag{3.232}$$

But, for a phase difference of $\pi/2$ (3.232) is nearly the same as the corresponding expression for a sphere of radius a and relative refractive index m. The equation (3.232) is correct to order $1/\rho^2$. Aas (1984) has examined the problem in the large ρ limit. In this limit the absorption and extinction efficiencies can be shown to be:

$$\bar{Q}^{abs}_{ADA} = 1 - \frac{3}{16(xn')^2}$$

and

$$\bar{Q}^{ext}_{ADA} \sim 2 - \frac{\cos\rho}{\rho} + \frac{3}{2\rho^2}\left(1 - \frac{\sin\rho}{4}\right). \tag{3.233}$$

It may be noted that equations (3.232) and (3.233) do not agree. The difference arises because – whereas in the first approach the averaging is done prior to taking the asymptotic limit – in the second approach the asymptotic limit is taken first and averaging is done later. Two operations do not commute.

The edge contribution to the extinction efficiency may be calculated from the formula (Jones, 1957):

$$Q_{edge} = \frac{c_0}{S}\int_P R^{1/3}\,ds, \tag{3.234}$$

where S is the projection of the scatterer on a plane perpendicular to the incident direction, R is the radius of curvature of the cylinder and s is the arc-length along the projection of the shadow boundary for a tilted cylinder. If Ψ is the angle between the axis of the cylinder and the direction of the incident wave, it is then possible to write the edge contribution as (Fournier and Evans, 1996):

$$Q_{edge} = \frac{c_o}{(x\sin\Psi)^{2/3}}.$$

The universal function c_0 is generally very complex, but for soft particles it can be approximated to be $c_0 = 0.996130$. Angle averaging then leads to:

$$\bar{Q}_{edge} \approx \frac{1.159595 c_0}{x^{2/3}}.$$

This edge contribution is valid only for large ρ_{ADA}. A semi-empirical modification that gives an edge correction that is valid for small ρ as well has also been obtained by Fournier and Evans (1996). It has been exhibited in the form:

$$\bar{Q}_{edge} = \frac{1.159\,59 c_0}{x^{2/3} + x_{crit}},$$

where x_{crit} is the size parameter that is approximately midway between 0 and the first peak. For a cylinder this gives:

$$x_{crit} = \frac{\pi}{4|m-1|}.$$

The approximate circular cylinder formula is far more economical in terms of computer resources than the exact, random, infinitely long cylinder code. It has been found that evaluation using an approximate formula is 10^4 times faster than with exact code. Typical errors are approximately 5% for $n \geq 1$ and $0 \leq n' \leq 3$ for medium- and large-size parameters.

Angle averaging over elliptic, infinitely long cylinders must be done over two angles, Ψ and ϕ. Here, ϕ is the angle between the semi-major axis of the ellipse and the plane defined by the direction of incident radiation and the infinite axis of the cylinder. Averaging over Ψ leads to:

$$\bar{Q}_{edge} = \frac{1.159\,59 c_0 r^{2/3}}{p^2 \left[b^{2/3} + \frac{\pi}{4|m-1|} \right]}, \quad (3.235)$$

where

$$p = (\cos^2 \phi + r^2 \sin^2 \phi)^{1/2} \quad (3.236)$$

and r is the ratio of the semi-major axis to semi-minor axis. The extinction efficiency $\bar{Q}_{ADA}^{ext}$ is the same as for circular cylinders, but with the replacement $x = rb/p$. Angle averaging over ϕ can be performed analytically, but leads to rather cumbersome expressions. It is, in fact, easier to perform integration over ϕ numerically.

3.9.2 Hexagonal columns

It has been suggested that a randomly oriented particle of volume V and projected area P is equivalent to a cylinder of the same volume with a thickness of V/P. The efficiency factors of this randomly oriented particle can be approximated by applying the ADA to the cylinder with incident radiation normal to its base (Bryant and Latimer, 1969). The theory has been referred to as the simplified anomalous diffraction theory (SADT). The ray-length is then the same for all the rays.

Consequently, the extinction and absorption efficiency factors of a randomly oriented particle can be expressed as (see, e.g., Sun and Fu, 2001):

$$Q_{SADT}^{ext} = 2 - 2\exp\left(-k\frac{V}{P}n'\right)\cos\left[k\frac{V}{P}(n-1)\right], \qquad (3.237)$$

and

$$Q_{SADT}^{abs} = 1 - \exp\left(-2k\frac{V}{P}n'\right). \qquad (3.238)$$

It is clear that Q_{SADT}^{ext} and Q_{SADT}^{abs} can be readily computed if V and P of a particle of arbitrary shape are known. The quantity V/P is referred to as the effective photon path. For randomly oriented hexagonal columns of base side a and height h, the quantity V/P is given by the formula:

$$\frac{V}{P} = \frac{2\sqrt{3}ah}{a\sqrt{3}+2h}. \qquad (3.239)$$

The predictions from (3.237) and (3.238) have been compared with computations from the standard ADA for spheres, randomly oriented finite circular cylinders and hexagonal columns. The refractive indexes used for comparison were $m = 1.1 + i0.01$, $1.10 + i0.10$ and $1.10 + i0.20$. The following conclusions have been derived from these studies.

(i) For spheres, the absorption efficiencies predicted by the SADT are greater in comparison to predictions of the ADA. The extinction curves are very different for weakly absorbing spheres, but agreement improves when absorption is stronger.
(ii) For finite cylinders of aspect ratio between 2 and 6, absorption efficiencies are significantly greater in the SADT than those predicted by the ADA. The difference can be about 15%. The difference between the SADT and the ADA for extinction efficiencies reaches up to 100%.
(iii) The results for hexagonal columns are similar to those for finite cylinders. The disagreement is slightly larger for hexagonal columns.
(iv) Since the ADA generally underestimates the true absorption efficiency, the SADT results may agree better with the exact results for certain size ranges.

The accuracy of the ADA in predicting the extinction efficiency and the single-scattering albedo for hexagonal column-like ice crystal particles has been examined by Liu et al. (1996). The rigorous results were obtained by the ray tracing method. The fractional error in extinction efficiency, defined as $(Q_{ex} - Q_{ADA})/Q_{ex}$, is found to be within 3%. The column length considered in this work was 120 μm and the radius was 30 μm. The incident wavelength ranged from 0.29 μm to 3.6 μm. This covers most solar radiation energy. The error in the ADA for single-scattering albedo has also been examined. When the wavelength is less than 1.2 μm, absorption is negligible and albedo is about 1. When absorption is not negligible, the errors are within 5% if the wavelength is less than 3.0 μm. For longer wavelengths the discrepancy is between 5%

and 10%. The parameters chosen are relevant to studies relating to cirrus clouds. ADA results for randomly oriented hexagonal ice crystals have been compared with exact calculations (FDTD) by Fu *et al.* (1999). It is noted that the absorption efficiency in the ADA generally underestimates the exact absorption efficiency, but approaches its correct value 1 for large-size parameters.

3.9.3 Finite cylinders

The reference to randomly oriented finite cylinders in the preceding section was restricted to a comparison between the SADT and the ADA. A comparison of the ADA with rigorous T-matrix computations of extinction and absorption efficiencies has been performed numerically by Liu *et al.* (1998). A typical illustration for absorption and extinction efficiencies has been made for $m = 1.0036 + i0.0923$ (ice) at wavelength $2.865\,\mu\text{m}$. Concurrence between T-matrix computations and ADA predictions is extremely good. Numerical validation shows that efficiencies in the ADA are not sensitive to the ratio $2a/L$ once the ratio is less than 0.1. On the basis of this observation it has been concluded that a cylinder can be accepted as an infinitely long cylinder if $2a/L < 0.1$.

Simple analytic constructions for extinction and absorption efficiencies can be obtained for the special case $|\rho^*| < 1$ (Aas 1984):

$$\bar{Q}_{ADA}^{ext} \simeq 2\kappa + \frac{4}{3}(\rho^2 - \kappa^2) - \frac{3}{2}\left(\rho^2\kappa - \frac{\kappa^3}{3}\right)\ln\left(\frac{L}{2a}\right), \tag{3.240}$$

$$\bar{Q}_{ADA}^{abs} \simeq 2\kappa - \frac{8}{3}\kappa^2 + 2\kappa^3 \ln\left(\frac{L}{2a}\right), \tag{3.241}$$

and

$$\bar{Q}_{ADA}^{sca} \simeq \frac{\pi^2}{8}(\rho^2 + \kappa^2) - \frac{3}{2}(\rho^2\kappa + \kappa^3)\ln\left(\frac{L}{2a}\right), \tag{3.242}$$

where L is the length of the cylinder. The results have been obtained by making suitable expansions in powers of ρ and κ.

3.9.4 Spheroids and ellipsoids

Angular averaging of the extinction efficiency factor for a spheroid can be carried out by employing the relation (Fournier and Evans, 1991):

$$\bar{Q}_{ext} = \frac{\int_0^{\pi/2} Q_{ext}\,\bar{p}\sin\Psi\,d\Psi}{\int_0^{\pi/2} \bar{p}\sin\Psi\,d\Psi}, \tag{3.243}$$

where $\bar{p}$ is defined as:

$$\bar{p} = \sqrt{\cos^2\Psi + r^2\sin^2\Psi} \tag{3.244}$$

A similar expression holds good for the absorption efficiency factor. It may be noted that for very large r, $\bar{p} = r \sin \Psi$ and (3.243) reduces to (3.230).

The extinction efficiency for randomly oriented spheroids has been examined numerically by Fournier and Evans (1991) in a soft-particle approximation referred to as the "Evans and Fournier approximation" (EFA). This approximation will be treated in detail in Chapter 4. For the time being, it suffices to give the outcome of the use of this approximation in this scattering problem. Numerical comparisons with rigorous computations have been performed in the range $1.01 \leq n \leq 2.0$, $0 \leq n' \leq 1$ and $0.1 \leq ka \leq 30$ (Fournier and Evans, 1991). It is found that for $r = 2$ the maximum error can go up to 30% (for $n < 1.02$). The error becomes considerably less ($<10\%$) for $n > 1.05$. Accuracy also increases with increasing absorption. This is true for all aspect ratios and sizes.

An alternative semi-empirical analytic expression for the extinction efficiency factor of randomly oriented spheroid has been obtained by Evans and Fournier (1994). The derivation is based on the EADA. Computations from the resulting formula have been contrasted with exact results. The approximate formula has been found to be good in the range $1 \leq n \leq \infty$, $0 \leq n' \leq \infty$ and $0.2 \leq r \leq 5.0$. It covers the entire domain of refractive index values. The approximate formula has been found to be uniformly good over all size parameters and aspect ratios.

In Section 3.7 it was noted that the integral radiative characteristics of an ellipsoid are equivalent to the integral characteristics of a sphere of radius h given by (3.192). The averaging procedure over all orientations can thus be achieved to a high degree of accuracy if one uses (Paramonov, 1994):

$$\langle h \rangle = \frac{3V}{2\langle P \rangle}, \qquad (3.245)$$

where $\langle P \rangle$ is the average value of the geometrical cross-section. Further, since $\langle P \rangle = S/2$ for randomly oriented convex particles, one can write $\langle h \rangle = 3V/S$ for randomly oriented ellipsoidal particles. Here, S is the surface area of the particle for a sphere $a = \langle h \rangle / 2$. This formula has been used successfully for many types of non-spherical particles including circular cylinders.

Shepelevich et al. (2001) have made an interesting observation regarding the equivalence between the radiative properties of a monodispersion of spheroids and a polydispersion of spheres. They have demonstrated that in the framework of the RGDA and the ADA, the light scattering characteristics of randomly oriented spheroids are equivalent to a polydispersion of spherical particles, the polydispersion being characterized by a power law size distribution. The upper and lower limits in this distribution correspond to maximal and minimal size parameters of the spheroid. An important characteristic of this distribution is that the average surface area and the volume of the spheres is equal to the surface area and volume of a spheroid.

On the question of equivalence, further work has been done within the framework of the Rayleigh approximation by Min et al. (2006). They have derived a theorem for the optical properties of particles of arbitrary shapes. The theorem is stated here without proof.

Theorem *The absorption cross-section of an arbitrarily shaped and arbitrarily oriented, homogeneous particle in the Rayleigh domain, or an ensemble of such particles with various shapes and orientations, equals the average cross-section of an ensemble of spheroidal particles in a fixed orientation with the same composition and a shape distribution that is independent of the composition of the particle.*

Evidently, this approach could be very useful for particles that are small compared with the wavelength of incident radiation. It may be noted that no limitation is placed on the refractive index of the particles.

3.9.5 Disks

Let Ψ be the angle between the symmetry axis of the disk and incident light direction. The geometrical cross-section of the disk is then $(\pi/4)D^2 \cos\Psi$ where D is the diameter of the disk. If L is the thickness of the disk, each ray passing through the disk travels a distance $L/\cos\Psi$. The mean extinction and absorption efficiencies of thin disks (diameter much greater than thickness) oriented at random can be expressed as (Aas, 1984):

$$Q_{ADA}^{ext} = 2 - 2\,\mathrm{Re}\left[e^{i\rho^*} + i\rho^* e^{i\rho^*} - \rho^{*2} E_1(-i\rho^*)\right], \qquad (3.246)$$

and

$$Q_{ADA}^{abs} = 1 - e^{-2\kappa} + 2\kappa e^{-2\kappa} - 4\kappa^2 E_1(2\kappa), \qquad (3.247)$$

where $a = L/2$. E_1 is the exponential integral and is given by relation (3.175). No numerical comparisons have been performed. Two asymptotic forms of extinction and absorption efficiency are:

$$Q_{ADA}^{ext} \simeq 2 - 4\,\mathrm{Re}\left[\frac{ie^{i\rho^*}}{\rho^*} + \frac{3e^{i\rho^*}}{\rho^{*2}}\right], \quad |\rho^*| \gg 1 \qquad (3.248)$$

and

$$Q_{ADA}^{abs} \simeq 1 - \frac{e^{-2\kappa}}{\kappa} + \frac{3e^{-2\kappa}}{2\kappa^2}, \quad \kappa \gg 1. \qquad (3.249)$$

The extinction efficiency oscillates around the value 2, and the absorption efficiency increases to 1 with increasing argument: the larger the imaginary part of the refractive index, the smaller the oscillations in the extinction curve. For nonabsorbing disks ($n' = 0$), equation (3.248) leads to:

$$Q_{ADA}^{ext} = 2 + \frac{4\sin\rho}{\rho} - \frac{12\cos\rho}{\rho^2} \qquad (3.250)$$

which strongly resembles the corresponding formula for spheres.

Derivation of the above formulas assumes that a negligible amount of light passes through the circular cylinder walls. The resulting efficiency factors are therefore independent of diameter D. When disk size decreases, the situation changes – especially when the axis of the disk is perpendicular to the direction of incidence.

The contribution from cylinder walls cannot then be neglected. A crude estimate then gives (Aas, 1984):

$$Q_{ext} \simeq 4\kappa + 2(\rho^2 - \kappa^2) \ln\left(\frac{D}{L}\right) \qquad (3.251)$$

$$Q_{abs} \simeq (4\kappa - 4\kappa^2) \ln\left(\frac{D}{L}\right), \qquad (3.252)$$

and

$$Q_{sca} \simeq 2(\rho^2 + \kappa^2) \ln\left(\frac{D}{L}\right). \qquad (3.253)$$

The efficiencies now depend on L as well as D.

3.10 POLYDISPERSION OF SPHERES

Analytic expressions can be obtained for ensemble-averaged absorption and extinction coefficients in the EA or ADA for certain size distributions. One such distribution – which is capable of representing cloud particles in the real atmosphere – is the gamma function and may be expressed as (Flatau, 1992):

$$n(a)\,da = \frac{N_t}{\Gamma(s)}\left(\frac{a}{a_p}\right)^{s-1} e^{-a/a_p}\,d(a/a_p), \qquad (3.254)$$

where a_p is the "characteristic radius", s is the measure of variance of the distribution, Γ is the gamma function and N_t is the total number of particles. The total area of particles S can be defined through the connection:

$$S = \int_0^\infty \pi a^2 n(a)\,da = s(s+1)N_t \pi a_p^2. \qquad (3.255)$$

This is the area of a characteristic particle of radius a_p multiplied by the total number of particles and adjusted by a constant $s(s+1)$. This example has been examined in detail by Flatau (1992).

If it is assumed that dispersion is tenuous, the particles in the ensemble can be regarded as independent scatterers and absorbers. The volume extinction and absorption coefficients for this problem may then be defined as:

$$\beta_{ADA}^{ext} = \int_0^\infty \pi a^2 n(a) Q_{ADA}^{ext}\,da, \qquad (3.256)$$

$$\beta_{ADA}^{abs} = \int_0^\infty \pi a^2 n(a) Q_{ADA}^{abs}\,da. \qquad (3.257)$$

For the distribution (3.254), the above equations lead to the following analytic expressions:

$$Q_{ADA}^{ext} = 4S\,\text{Re}\,\mathcal{M}(\omega_p^*), \qquad (3.258)$$

$$Q_{ADA}^{abs} = 2S\mathcal{M}(4\kappa_p), \qquad (3.259)$$

where
$$\omega_p^* = 2ix_p(n-1) - 2x_p n'$$
with
$$x_p = ka_p, \quad \kappa_p = x_p n'$$
and the function:
$$\mathcal{M}(\omega^*) = \frac{1}{2} + \frac{1}{s(1+s)}\left[-\frac{s}{\omega^*(1+\omega^*)^{1+s}} + \frac{1}{\omega^{*2}(1+\omega^*)^s} - \frac{1}{\omega^{*2}}\right], \quad (3.260)$$

The function $\mathcal{M}(\omega^*)$ asymptotes to $1/2$ for large ω^* and to 0 for small ω^*. These asymptotic values are the same as those for single-particle scattering $\mathcal{K}(\omega)$. They have a similar behavior but appear to be shifted with respect to each other.

From (3.258) and (3.259), the distribution-averaged single-scattering albedo in the ADA is:
$$\tilde{\omega}_{ADA} = 1 - \frac{2\mathcal{M}(2\kappa_p)}{4 \operatorname{Re} \mathcal{M}(2\omega_p)}, \quad (3.261)$$

which for the particular case of $\beta_{sca} = 2S$ gives:
$$\tilde{\omega}_{ADA} = \frac{1}{2} - \frac{1}{(s+1)s}\left[\frac{s}{\kappa_p(\kappa_p+1)^{s+1}} + \frac{1}{\kappa_p^2(\kappa_p+1)^s} - \frac{1}{\kappa_p^2}\right]. \quad (3.262)$$

As expected, it gives $\tilde{\omega}_{ADA} = 1$ for $n' = 0$.

This is just one example that shows the amount of simplification that can be achieved by the EA or the ADA in treating the complicated problem of scattering and absorption of radiation by an ensemble of polydisperse particles. The precision of the anomalous diffraction theory has been studied by Ackerman and Stephans (1987) for a collection of spherical particles where size distribution is represented by the modified gamma distribution. Analytic expressions have been obtained by them for extinction and absorption efficiency factors for a number of size distributions. However, we do not reproduce them here. They are somewhat more cumbersome in appearance than those obtained above. Here, we only give some results from the numerical comparisons for the modified gamma distribution. Errors for a size distribution are found to be much smaller than those for a single particle. For cloud water droplets where $\beta \leq 0.2$, errors are typically less than 10% for the extinction efficiency factor and less than 30% for the absorption efficiency factor. Analytic expressions for absorption efficiencies, scattered efficiencies and for the single-scattering albedo for polydispersions of horizontally oriented hexagonal columns and hexagonal plates have been derived by Chýlek and Videen (1994) in the ADA for a gamma-type size distribution. The derivations are straightforward. Numerically, ADA predictions have been contrasted with discrete dipole approximation computations. The error estimates (deviation from the DDA) show that the accuracy of ADA predictions lie within 5% in the bands 2.77 to 3.02 µm and from 10.6 to 11.1 µm. The errors were within 15% in the bands 0.8–3.05, 5.4–6.1 and 8.3–11.9 µm. Errors due to the use of equivalent spheres in the calculation of various quantities were typically of the order of several hundred percent.

4

Other soft-particle approximations

In this chapter we turn our attention to soft-particle approximations other than the eikonal approximation (EA) or the anomalous diffraction approximation (ADA). These approximations include the Perelman approximation (PA), the Hart and Montroll approximation (HMA), the Evans and Fournier approximation (EFA), the Bohren and Nevitt approximation (BNA), the Nussenzveig and Wiscombe approximation (NWA) and the Pendorf–Shifrin—Punina approximation (PSPA). The Rayleigh–Gans–Debye approximation (RGDA) has been dealt in detail in many books and articles. Here, we include a brief recap of this approximation for reference and completeness of discussion.

4.1 RAYLEIGH–GANS–DEBYE APPROXIMATION

In the literature, various permutations and combinations of these three names have been used to refer to this approximation. The papers generally associated are Rayleigh (1881, 1914), Gans (1925) and Debye (1915). In addition to these names, citation has also been made to the Rayleigh–Gans–Born approximation (Turner, 1976) and the Rayleigh–Gans-Rokard approximation (Acquista, 1976). In this book we refer to this approximation as the Rayleigh–Gans–Debye approximation (RGDA). This simplification of rigorous solutions has been designed to work in the domain:

$$|m - 1| \ll 1, \qquad (4.1)$$

and

$$x|m - 1| \ll 1. \qquad (4.2)$$

Since $|m - 1| \ll 1$, it may be argued that reflection at the boundaries can be ignored. Further, as the strength of the interaction $x|m - 1|$ is also small, the electric field inside the particle may be taken to be the same as that of the incident wave. Let us denote by $S_1(\theta)$ the scattering function when the electric vector is perpendicular to the plane of scattering and by $S_2(\theta)$ the scattering function when the electric vector is parallel to the plane of scattering. The RGDA leads to the following scattering matrix for amplitudes $S_1(\theta)$ and $S_2(\theta)$:

$$\begin{pmatrix} S_1 \\ S_2 \end{pmatrix} = -ik^3 \alpha \begin{pmatrix} 1 \\ \cos\theta \end{pmatrix} R(\theta, \varphi), \qquad (4.3)$$

where $\alpha = (m^2 - 1)V/(4\pi) \approx (m-1)V/2\pi$ is the polarizability of the particle which is assumed to be isotropic, V is the volume of the obstacle and the shape factor $R(\theta, \varphi)$ is:

$$R(\theta, \varphi) = \frac{1}{V} \int_V e^{i\Delta(\mathbf{r})} \, dV, \qquad (4.4)$$

with $\Delta(\mathbf{r})$ as the phase difference between the waves arriving in a given direction after scattering by various volume elements in the scatterer. The integral in $R(\theta, \varphi)$ can be evaluated for most simple shapes. A number of authors (van de Hulst, 1957; Kerker, 1969; Bayvel and Jones, 1981; and Kokhanovsky, 2005) give values of $R(\theta, \varphi)$ for some particle shapes. However, for an arbitrary shape it is not easy to evaluate $R(\theta, \varphi)$. In what follows we restrict ourselves to two simple cases – namely, a homogeneous sphere and an infinitely long circular cylinder.

4.1.1 Homogeneous sphere

For a homogeneous sphere the polarizability and the shape factor can be easily calculated and the scattered intensity for unpolarized light can be written as:

$$\begin{aligned} i(\theta)_{RGDA} &= \tfrac{1}{2}[|S_1(\theta)|^2 + |S_2(\theta)|^2] \\ &= \tfrac{2}{9} x^6 |m-1|^2 (1 + \cos^2\theta) |R(\theta)|^2, \end{aligned} \qquad (4.5)$$

where $S_1(\theta)$ and $S_2(\theta)$ are as defined in (4.3). The shape factor $R(\theta)$ takes a simple form:

$$R(\theta) = \frac{3}{z^3}(\sin z - z\cos z). \qquad (4.6)$$

As before, $z = 2x\sin(\theta/2)$. It may be noted that for a nonabsorbing particle the amplitudes given by (4.3) are purely imaginary. The optical theorem then gives:

$$Q_{RGDA}^{ext} = \frac{4}{x^2} \operatorname{Re} S(0) = 0. \qquad (4.7)$$

Clearly, this is not a correct result. It implies that the optical theorem cannot be used to calculate the extinction efficiency factor on the basis of the RGDA amplitude. The reason being that the RGDA amplitude does not satisfy the unitarity property. The extinction efficiency in the RGDA, therefore, should be calculated as a sum of Q_{RGDA}^{abs} and Q_{RGDA}^{sca}.

The rigorous scattering efficiency defined as:

$$Q_{sca} = \frac{1}{x^2} \int (|S_1(\theta)|^2 + |S_2(\theta)|^2) \sin\theta \, d\theta, \tag{4.8}$$

can be easily calculated using (4.5) and (4.6). This gives:

$$Q_{RGDA}^{sca} = |m-1|^2 \zeta(x), \tag{4.9}$$

where

$$\zeta(x) = \left[\frac{5}{2} + 2x^2 - \frac{\sin 4x}{4x} - \frac{7}{16x^2}(1-\cos 4x) + \left(\frac{1}{2x^2} - 2\right)(\gamma + \log 4x - \text{Ci}(4x))\right], \tag{4.10}$$

with Ci(x) as the cosine integral given by (3.172) and $\gamma = 0.577$ as the Euler's constant. The absorption efficiency factor in the RGDA gives:

$$Q_{RGDA}^{abs} = \frac{8x}{3} \text{Im}(m-1), \tag{4.11}$$

and Q_{RGDA}^{ext} is the sum of (4.9) and (4.11). For $\rho^* = 2x(m-1) \ll 1$, the errors are reasonably small. This approximation for extinction efficiency in the domain $1.0 < m \leq 1.25$ is better than 10% for $x < 1$ (Bayvel and Jones, 1981). For a fixed value of m, accuracy decreases as x increases. If $x \gg 1$, (4.9) gives:

$$Q_{RGDA}^{sca} = 2x^2 |m-1|^2. \tag{4.12}$$

This simplification is known as Jobst's approximation (Jobst, 1925).

Equation (4.6) tells us that the minima of the scattering pattern for a sphere are determined by the zeros of the equation $\tan z = z$. That is, the zeros in the scattering pattern are dependent only on the size of the scatterer and not on its refractive index. To a good approximation the nth zero is then given by the relation $z_n^2 = (n+1/2)^2 \pi^2 - 2$, $n = 1, 2, \ldots$. There are, however, no zeros of $R(\theta)$ for any θ unless $x > 2.25$.

Shimizu (1983) and Gordon (1985) have examined modifications to the RGDA. The underlying idea in both the modifications is to introduce refractive index dependence in $R(\theta)$. While the modified version of Shimizu changes x to mx, Gordon argues for the substitution $x(1 + m^2 - 2m\cos\theta)^{1/2}$ in place of $2x\sin(\theta/2)$. The result of RGDA is based on the assumption that the field inside the particle can be taken to be the same as the incident field. The modified RGDA of Gordon corrects this simplification by arguing that – while the field inside the scattering region is still a plane wave – the wave number changes. The new wave number is a product of the incident wave number and the refractive index of the obstacle.

Modified versions give considerably better agreement with the Mie theory for positions of extrema in the scattering pattern. However, the magnitude of scattered intensity at the minima is quite different from exact results. This discrepancy can be removed by adding an x-dependent function $\gamma(x)$ to $R(\theta)$ (Gordon, 1985). The new

$R(\theta)$ is then defined as:

$$R(\theta) = \frac{3(\sin z - z \sin z)}{z^3} + \gamma(x),\qquad(4.13)$$

where

$$z = x(1 + m^2 - 2m \cos\theta)^{1/2}.$$

For a homogeneous sphere, it is found that:

$$\gamma(x) = x^{-3/2},$$

provides reasonable agreement with Mie theory up to about $x = 30$ for the refractive index range $1.0 < m < 1.2$. Note that for $m = 1.2$ and $x = 30$ we have $\rho = 12$. Clearly, despite the original premise that $\rho < 1$ for the validity of the RGDA and its modified variants, these give good results even when ρ is considerably greater than 1.

Another problem with the RGDA that needs to be addressed is the fact that it does not work well for shapes which differ strongly from spherical geometry (see, e.g., Barber and Wang, 1978) for $\rho \geq 1$. To overcome this difficulty an alternative known as the quasi-static approximation (QSA) was developed for cylinders by Burberg (1956) and for spheroids by Shatilov (1960). More recently, Voshchinnikov and Farafonov (2000), Farafonov et al. (2001) and Posselt et al. (2002) have further examined QSA for spheroids, infinite cylinders and multi-layered spheroids. In this approximation the field inside the scatterer is represented by the incident plane wave (as in the RGDA), while the polarizability of the particle is taken into account as in the Rayleigh approximation. In this sense QSA is a generalization of the Rayleigh approximation and the RGDA. The condition for applicability of the QSA is:

$$|m - 1|x_v \ll 1,$$

where, for a spheroid, $x_v = 2\pi a_v/\lambda$ (a_v is the radius of a sphere whose volume is equal to that of the spheroid). Numerical results show that the QSA works particularly well for small optically soft particles. For such particles it is always preferable to the RGDA. It is found that – while the range of applicability of RGDA decreases with growing asphericity – the validity range of QSA remains practically independent of scatterer shape. In comparison to the RGDA, its range of applicability is nearly always greater if $a/b \geq 3$. Here a and b are the semi-major and semi-minor axes, respectively. The above discussion essentially pertains to the scattering efficiency factor. The positions of minima, however, continue to be reproduced correctly even as the deviation from sphericity increases (Barber and Wang, 1978).

In an ensemble of particles the effect of nonsphericity is not significant for small particles. If the particle size is such that the Rayleigh approximation is applicable, it is found that the light scattered by randomly oriented spheroids is almost indistinguishable from light scattered by a collection of spheres of equal volume (Paramonov et al., 1986).

A series representation of $\zeta(x)$, particularly useful for polydispersions, has been given by Shifrin and Ston (1976):

$$\zeta(x) = \sum_{l=1}^{\infty} (-1)^{l+1} \frac{(4x)^{2l+2}}{(2l+2)} \frac{l^2+l+2}{2(l+2)^2(l+1)}. \qquad (4.14)$$

Further, it has been shown that the following approximations:

$$\zeta(x) = \begin{cases} 1.185x^4(1.0 - 0.4x^2 + 0.096x^4) & x \in [0,1] \\ 1.92x^4 - 1.084x & x \in [1,2] \\ 2.112x^2 - 1.456x & x \in [2, 12.5], \end{cases} \qquad (4.15)$$

derived from (4.14) can be used to within 3% accuracy.

An interpolation formula for a nonabsorbing sphere, based on RGDA, has been obtained by Walstra (1964). If, for a given x, Q_{ext} is known for $m = a$ and $m = c$, Q_{ext} can be computed for the intermediate value of $m = b$ using the relation:

$$Q_{ext}(b) = \frac{(b-1)^2}{c-a} \left[\frac{c-b}{(a-1)^2} Q_{ext}(a) + \frac{b-a}{(c-1)^2} Q_{ext}(c) \right]. \qquad (4.16)$$

The results from this formula yield less than a 1% error for values of x up to at least 8 in the refractive index range $m = 1.025(0.05)1.275$. Walstra (1964) has also given simple empirical formulas for the extinction efficiency factor. For small particles it has the form:

$$Q_{WAS}^{ext} = (1.26m - 0.04)\rho - 2.558(m-1)^{1.273} - 0.843.$$

This gives less than a 1% error in the range $1.5 < \rho < 2.5$.

The asymmetry parameter takes the following simple form in the RGDA (Kokhanovsky and Zege, 1997, Irvine, 1963):

$$g_{RGA} = \frac{\zeta(x)}{H(x)} \qquad (4.17)$$

where

$$H(x) = \frac{4}{x^4} \left[\left(\frac{9}{128} - \frac{5x^2}{64} \right) (4x \sin 4x + \cos 4x) + \frac{x^6}{2} + \frac{11x^4}{8} \right. $$
$$\left. - \frac{31x^2}{64} - \frac{9}{128} + x^2 \left(x^2 - \frac{3}{8} \right) (\text{Ci}(4x) - \gamma - \ln(4x)) \right]. \qquad (4.18)$$

It may be noted that the asymmetry parameter is independent of refractive index in this approximation.

4.1.2 Scattering by an infinitely long cylinder

If the radiation is incident on an infinitely long cylinder of radius a at an angle perpendicular to the axis of the cylinder, the scattering functions in the RGDA can be written as (see, e.g., Bohren and Huffman, 1983 or van de Hulst, 1957):

$$T(\theta)_{RGDA}^{TMWS} = \frac{-i\pi(m-1)x^2 J_1(z)}{z}, \qquad (4.19)$$

and

$$T(\theta)_{RGDA}^{TEWS} = \cos\theta\, T(\theta)_{RGDA}^{TMWS}. \qquad (4.20)$$

The corresponding formula for finite cylinders can be found in Bohren and Huffman (1983) and Van de Hulst (1957). Modifications analogous to those described by Shimizu (1983) and Gordon (1985) for a sphere have been studied by Sharma and Somerford (1988) for an infinitely long cylinder for normal incidence. These are found to reproduce the positions of extrema much more accurately than the RGDA. Detailed numerical results for deducing the diameter of a circular cylinder on the basis of positions of minima are given in Chapter 5.

The large diameter limit of the scattering efficiency factor for a nonabsorbing infinitely long cylinder may be obtained by employing the extinction theorem $Q_{ext} = (2/x)\,\text{Re}\,T(0)$. This gives:

$$Q_{RGDA}^{sca}(\text{cylinder}) = \tfrac{8}{3}k^2|m-1|^2 a^2.$$

The above result is identical to that obtained from the ADA in the small $|\rho_{ADA}^*|$ limit. However, a similar equivalence does not hold for scattered intensity. Recall that for scattered intensity the ADA is equivalent to the RGDA only for the small $|\rho_{ADA}^*|$ limit and not to its large size limit.

4.2 PERELMAN APPROXIMATION

Perelman (1978, 1991) showed that it is possible to sum the Mie series for small-angle scattering by executing the limit $m \to 1$ in the denominators of the scattering coefficients. The summation leads to two closed form analytic approximations: (*i*) the simple form of the Perelman approximation (PA) and (*ii*) the main form of the Perelman approximation (MPA).

4.2.1 Homogeneous sphere

A convenient way to introduce the PA consists in first rewriting the Mie scattering coefficients (3.54) and (3.55) as:

$$a_l = \frac{h_{1l}}{h_{1l} + ih_{3l}}, \qquad (4.21)$$

and

$$b_l = \frac{h_{2l}}{h_{2l} + ih_{4l}}, \qquad (4.22)$$

where

$$h_{1l} = mu_l(mx)u_l'(x) - u_l'(mx)u_l(x),$$
$$h_{2l} = u_l(mx)u_l'(x) - mu_l'(mx)u_l(x),$$
$$h_{3l} = mu_l(mx)v_l'(x) - u_l'(mx)v_l(x),$$
$$h_{4l} = u_l(mx)v_l'(x) - mu_l'(mx)v_l(x),$$
(4.23)

with $u_l(x) = xj_l(x)$ and $v_l(x) = xn_l(x)$ as the Riccati–Bessel functions. The spherical Bessel function j_l and spherical Neumann function n_l are related to the corresponding cylindrical functions J_l and N_l via the relations:

$$j_l(x) = \sqrt{\frac{\pi}{2x}} J_{l+1/2}(x), \quad \text{and} \quad n_l(x) = \sqrt{\frac{\pi}{2x}} N_{l+1/2}(x). \qquad (4.24)$$

It can then easily be ascertained that in the limit $m \to 1$ the denominators in (4.21) and (4.22) can be simplified to:

$$h_{1l} + ih_{3l} \sim h_{2l} + ih_{4l} = i|m|^{-1/2}. \qquad (4.25)$$

The Mie scattering coefficients a_l and b_l then become:

$$a_l = \frac{i}{|m|^{1/2}} h_{1l}, \qquad (4.26)$$

and

$$b_l = \frac{i}{|m|^{1/2}} h_{2l}. \qquad (4.27)$$

The summation over l in equations (3.52) and (3.53) for scattering functions $S_1(\theta)$ and $S_2(\theta)$ can now be carried out readily for near-forward scattering. Expanding $\pi_l(\theta)$ and $\tau_l(\theta)$ for small θ and using the addition theorems given in Appendix B, approximate scattering functions may be expressed as:

$$S_1(\theta)_{PA} = -2i(m+1)m|m|^{-1/2} x^2 \left[\frac{u_1(\rho/2)}{\rho} - 2(1-\mu)mx^2 \frac{u_2(\rho/2)}{\rho^2} \right], \qquad (4.28)$$

and

$$S_2(\theta)_{PA} = -2i(m+1)m|m|^{-1/2} x^2 \left[\frac{\mu u_1(\rho/2)}{\rho} - 2(1-\mu)mx^2 \frac{u_2(\rho/2)}{\rho^2} \right], \qquad (4.29)$$

where $\mu = \cos\theta$ and $\rho = 2x(m-1)$. The scattering functions (4.28) and (4.29) are correct to order θ^4 as $\theta \to 0$. In other words, the terms neglected in expansions of $\pi_l(\theta)$ and $\tau_l(\theta)$ are of order higher than θ^4.

For small particles (in the limit $x \to 0$, $\rho \to 0$), one may approximate $u_1(\rho/2)$ and $u_2(\rho/2)$ by the first term in the expansions. Thus, one may write:

$$u_1(\rho/2) = \frac{\rho^2}{12} + O(\rho^4); \quad u_2(\rho/2) = \frac{\rho^3}{120} + O(\rho^5).$$

The first terms in the square brackets in (4.28) and (4.29) are the leading terms in this limit. The equations in this limit, therefore, give:

$$S_1(\theta)_{PA} \approx \frac{-im|m|^{-1/2}}{3}(m^2-1)x^3, \qquad (4.30)$$

and

$$S_2(\theta)_{PA} \approx \cos\theta \, S_1(\theta). \qquad (4.31)$$

It can be seen that $S_1(\theta)_{PA}$ and $S_2(\theta)_{PA}$ are nothing more than the scattering functions in the Rayleigh approximation for soft particles. As one is considering the $m \to 1$ approximation, $3|m|^{1/2}$ may be replaced by $m(m^2+2)$.

In the RGDA, the validity domain is $\rho < 1$ and $|m-1| \ll 1$. Both the terms in square brackets (4.28) and (4.29) now need to be retained. The scattering function with a small-angle approximation can then be cast in the form:

$$S_1(\theta)_{PA} = \frac{-im|m|^{1/2}}{3}(m^2-1)x^3, \qquad (4.32)$$

$$S_2(\theta)_{PA} = \frac{-im|m|^{1/2}}{3}(m^2-1)x^3 \left[\mu - \frac{mx^2}{5}(1-\mu)\right]. \qquad (4.33)$$

The equivalence of the above equations with the corresponding RGDA expressions given by (4.5) becomes transparent when $R(\theta) = (\sin z - z\sin z)/z^3$ given in (4.6) is expanded in powers of θ. To the order θ^4, the scattering functions in the RGDA are found to be the same as the scattering functions (4.32) and (4.33).

The main form of the PA is obtained by recasting scattering coefficients in the form:

$$a_l = \frac{h_{1l}(h_{1l} - ih_{3l})}{h_{1l}^2 + h_{3l}^2}, \qquad (4.34)$$

and

$$b_l = \frac{h_{2l}(h_{2l} - ih_{4l})}{h_{2l}^2 + h_{4l}^2}. \qquad (4.35)$$

This is achieved by multiplying and dividing a_l and b_l – given by (4.21) and (4.22) – by the complex conjugates of respective denominators. The MPA then consists in approximating:

$$h_{1l}^2 + h_{3l}^2 \sim h_{2l}^2 + h_{4l}^2 = |m|. \qquad (4.36)$$

The resulting series can be summed for forward scattering to yield (Perelman, 1991):

$$S_1(0)_{MPA} = S_2(0)_{MPA}$$

$$S(0)_{MPA} = \frac{x^2}{8|m|}\left[(m^2+1)^2 + \frac{w(m,\rho) - w(-m,-R)}{2m}\right], \qquad (4.37)$$

where
$$w(m, z) = [a(m) + a_0(m)z^2]ei(z) - ia_1(m)e_1(z) + a_2(m)e_2(z),$$
$$a(m) = (m^2 - 1)^2(m^2 + 1), \quad a_0(m) = -2(m^2 - 1)^2(m - 1)^2,$$
$$a_1 = (m + 1)^2(m^4 - 2m^3 - 2m^2 - 2m + 1),$$
$$a_2(m) = -a_0(m) - a_1(m),$$
$$ei(z) = \int_0^z dt\,(1 - \exp(-it))/t, \quad e_1(z) = \exp(-iz)/z,$$
$$e_2(z) = (1 - \exp(-iz))/z^2,$$

and $R = 2x(m + 1)$. The addition theorems necessary for summation are given in Appendix B. The above result holds for all values of $x > 0$ and $m = n + in'$. The extinction efficiency factor can be obtained from (4.37) using the optical theorem (3.7).

The Mie series can be summed for non-forward scattering too, but only for small scattering angles. The result is (Perelman 1991):

$$S_k(\mu)_{MPA} = S(0)_{MPA} - (1 - \mu)H_k; \quad k = 1, 2, \qquad (4.38)$$

where
$$H_1 = -i(m + 1)m^2|m|^{-1/2}x^4z^2 u_2(z), \qquad (4.39)$$

and
$$H_2 = -i(m + 1)m^2|m|^{-1/2}x^4\left[z^2\psi_2(z) + \frac{x^2}{mz}u_1(z)\right]. \qquad (4.40)$$

In (4.38) the scattering functions are correct to order θ^4.

The extinction efficiency factor can be calculated either by first inserting the approximation in the amplitude and then following it using the optical theorem or by first writing the exact analytic form of the extinction efficiency and subsequently implementing the approximation (4.36) in it. The second approach is generally preferable because one is then not unduly worried about questions relating to the unitarity property. For a homogeneous sphere, the extinction efficiency can be written as,:

$$Q^{ext}_{MIE} = \frac{4}{x^2}\operatorname{Re} S(0) = \frac{2}{x^2}\operatorname{Re}\sum_{l=1}^{l=\infty}(2l + 1)[a_l + b_l], \qquad (4.41)$$

where the relations $\pi_l(0) = \tau_l(0) = l(l + 1)/2$ have been used. Perelman (1978), starting from (4.41), and employing the simplification (4.36) has obtained the following expression for the extinction efficiency factor of a nonabsorbing spherical particle:

$$Q^{ext}_{MPA} = b_1\left[Q_h(\rho) - \left(\frac{m-1}{m+1}\right)^2 Q_h(R)\right] + b_2\left[Q_1(\rho) - Q_1(R) - \int_\rho^R \frac{1 - \cos t}{t}dt\right]$$
$$+ \frac{(m-1)^2}{4m^2x^2}\left[Q_2(R) - Q_2(\rho) + \frac{1}{2}\int_\rho^R \frac{1 - \cos t}{t}dt\right] + \frac{1}{2mx^2}Q_3(m, x), \qquad (4.42)$$

where

$$b_1 = \frac{(m+1)^2(m^4+6m^2+1)}{32m^2}, \quad b_2 = \frac{(m^2+1)(m^2-1)^2}{4m^2}$$

$$Q_1(\omega) = \frac{2(1-\cos\omega)}{\omega^2} + \frac{2\sin\omega}{\omega},$$

$$Q_2(\omega) = \frac{3\cos\omega}{8} + \frac{\sin\omega}{8} + \frac{\omega^2}{16}$$

$$Q_3(\omega) = (m-1)^2(\cos R - 1) + (m+1)^2(\cos\rho - 1),$$

and

$$Q_h(\rho) = 2 - 4\frac{\sin\rho}{\rho} + 4\frac{1-\cos\rho}{\rho}. \tag{4.43}$$

The extinction efficiency can also be cast in an integral representation. For a non-absorbing spherical particle, Granovskii and Ston (1994a, b) have obtained the following simple-looking formula:

$$Q_{MPA}^{ext} = mx^4(m^2-1)^2 \int_{-1}^{1} dt\, (1+t^2)g^2(x\omega(t)), \tag{4.44}$$

where

$$g(\omega) = (\omega\cos\omega - \sin\omega)/\omega^3 \tag{4.45}$$

and

$$\omega(t) = (1+m^2-2mt)^{1/2}. \tag{4.46}$$

It can readily be verified by substituting $m = 1$ in (4.46) that equation (4.44) can be further approximated as:

$$Q_{MPA}^{ext} = 4x^4(m-1)^2 \int_{-1}^{1} dt\, (1+t^2)g^2(x\sqrt{2(1-t)}), \tag{4.47}$$

Equation (4.47) is nothing more than the integral representation of the extinction efficiency in the RGDA (Granovskii and Ston, 1994a, b). Integral representations of the amplitude function and Fourier spectra of components of electromagnetic field vectors have also been obtained by Perelman (1997) in this approximation.

The extinction efficiency factors computed using the main form of the PA agree with exact results to within 5% in the refractive index range $1.1 \leq m \leq 1.5$ and size parameter range $0 \leq x \leq x(m)$, if:

$$x(m) \approx 290m^{-11}, \tag{4.48}$$

where $x(m)$ is the upper limit of x for a given m for which errors are within 5% of exact results.

Perelman and Voshchinnikov (2002) have considered corrections to the main form of the PA for particles with real refractive index. For this purpose, they re-write approximation (4.36) as:

$$h_{1l}^2 + h_{3l}^2 \approx h_{2l}^2 + h_{4l}^2 \approx |m|/\tilde{g}(m,x), \tag{4.49}$$

where $\tilde{g}(m, x)$ represents the correction with the condition that $g(m, x) = \mathcal{O}(1)$ as $x \to \infty$. By examining the short-wavelength asymptotics of the extinction efficiency factor it can be shown that:

$$S(m) = \lim_{x \to \infty} Q_{MPA}^{ext}$$
$$= \frac{(m2+1)^2}{2m} + \frac{(m^2+1)^2(m^2-1)^2}{4m^2} \ln\left|\frac{m-1}{m+1}\right|, \qquad (4.50)$$

and then $\tilde{g}(m, x)$ is found to be:

$$\tilde{g}(m, x) = 1 - \tilde{f}(m, x) + \frac{2}{S(m)} \tilde{f}(m, x), \qquad (4.51)$$

where

$$\tilde{f}(m, x) = \exp\left[-\frac{0.01 \exp(4m) x(m)}{x}\right]. \qquad (4.52)$$

The extinction efficiency with this improvement can be written as:

$$Q_{IPA}^{ext} = \left[1 - \frac{S(m)-2}{S(m)} \exp\left(-\frac{0.01 \exp(4m)}{u}\right)\right] Q_{MPA}^{ext}, \qquad (4.53)$$

where $u = x/x(m)$ is the size parameter in the scale of $x(m)$ depending on m. The subscript *IPA* stands for improved Perelman approximation.

For absorption efficiency, Perelman (1994) has been able to re-cast the series in such a way that it rapidly converges for weakly absorbing particles. The end result of a lengthy calculation for a soft sphere is:

$$Q_{MPA}^{abs} = 4n|m|^2 S(\kappa), \qquad (4.54)$$

where

$$S(\kappa) = \frac{\tau \cosh \kappa - \sinh \kappa}{\kappa^2}, \qquad (4.55)$$

$m = n + in'$ is the relative refractive index of the particle and, as defined earlier, $\kappa = 2n'x$. The approximate formula is valid only for $\kappa \ll 1$.

4.2.2 Special cases

If $x \ll 1$, the basic formula (4.42) can be expanded in powers of x. It can easily be verified that:

$$Q_{MPA}^{ext} = \frac{8}{27} m(m^2-1)^2 x^4. \qquad (4.56)$$

This differs from the Rayleigh formula only in that $(m^2+1)^2$ in the original Rayleigh formula is replaced by 9. For a soft particle $m \to 1$ and this change should not cause an appreciable difference in the outcome.

Let us now examine equation (4.42) under the condition $|m-1| \ll 1, \rho \ll 1$. One can put $b_2 = 0$ and, hence, the first term in Q_3 is also 0. The dependence of other terms on ρ may be specified as follows:

$$Q_h(\rho) = \frac{(m-1)^2 R^2}{8}, \quad Q_1 = 3, \quad \int_0^\rho dt (1 - \cos t)/t = 0,$$

and

$$Q_2(\rho) = \frac{(m-1)^2}{8},$$

if terms of order higher than $(m-1)^2$ are neglected. Equation (4.42) then reduces to:

$$Q_{MPA}^{ext} = (m-1)^2 \left[\frac{5}{2} + 2x^2 - \frac{\sin 4x}{4x} - \frac{7}{16x^2}(1 - \cos 4x) + \left(\frac{1}{2x^2} - 2 \right)(\gamma + \log 4x - \text{Ci}(4x)) \right],$$

which is nothing more than Q_{RGDA}^{ext} given in (4.9). This is as it should be, because $(m-1) \ll 1$ and $\rho \ll 1$ constitute the domains where the RGDA is expected to yield good results.

For $m \to 1$, $x \to \infty$ but ρ is finite, various terms of Q_{ext}^{MPA} are of the order:

$$b_1 \left(\frac{m-1}{m+1} \right)^2 Q_h(R) = O(m-1)^2,$$

$$b_2[Q_1(\rho) - Q_1(R)] = O(m-1)/x,$$

$$b_2 \int_\rho^R \frac{1 - \cos t}{t} dt = O(m-1)\rho,$$

$$\frac{(m^2-1)^2}{4m^2 x^2}[Q_2(R) - Q_2(\rho)] = O(m-1)^2,$$

$$\frac{(m^2-1)^2}{8m^2 x^2} \int_\rho^R \frac{1 - \cos t}{t} dt = O(m-1)^2/x,$$

$$\frac{1}{2mx^2} Q_3(m,x) = O(x^{-2}).$$

Clearly, the third of these terms is the leading contributor. Equation (4.42) then gives (Perelman, 1978):

$$Q_{MPA}^{ext} = b_1 Q_h(\rho) - b_2 \int_\rho^R \frac{1 - \cos t}{t} dt. \tag{4.57}$$

It has been shown that (4.57) results in an improvement over the ADA extinction coefficient (Perelman, 1978). Further, for large values of x, the cosine integral can be replaced by its asymptotic value:

$$\int_\rho^R \frac{1 - \cos t}{t} dt = \ln \frac{R}{\rho} - \left. \frac{\sin t}{t} \right|_\rho^R + \int_\rho^R \frac{\sin t}{t} dt \approx \ln \frac{m+1}{m-1}.$$

The second term on the right-hand side of equation (4.57) is, therefore, of relative order $(m-1)$ and may be ignored. Thus, if one is considering the asymptotic value of the size parameter, Q_{MPA}^{ext} is related to Q_h via a multiplicative factor b_1.

4.2.3 Backscattering

In the PA and the MPA, the Mie series can be summed to yield simple expressions for backscattering too (Perelman, 1985). In this event the scattering amplitudes simplify to:

$$S_1(\pi) = S_2(\pi) \equiv S(\pi) = \frac{1}{2}\sum_{l=1}^{\infty}(2l+1)(-1)^l[a_l - b_l]. \quad (4.58)$$

Invoking approximation (4.25), the series in (4.58) can be summed to yield the following simple expression:

$$S(\pi)_{PA} = \frac{x|m-1|\sqrt{m}}{m+1} u_1((m+1)x), \quad (4.59)$$

where $u_l(x) = xj_l(x)$ is the Riccati–Bessel function. Further, using the approximation (4.36) in place of (4.25), one gets:

$$S(\pi)_{MPA} = \frac{\rho_{EA}^2}{2\pi}\left[\int_0^1 z(t)\,dt + s((m-1)x)s((m+1)x)\right], \quad (4.60)$$

where

$$z(t) = -z_0(t) + mx^2 t z_1(t) + m^2 x^4(1+t^2)z_2,$$

with

$$z_0(t) = \frac{u_0(\omega)u_0(\bar{\omega})}{\omega\bar{\omega}},$$

$$z_1(t) = \frac{u_0(\omega)u_1(\bar{\omega})}{\omega\bar{\omega}^2} - \frac{u_1(\omega)u_0(\bar{\omega})}{\omega^2\bar{\omega}},$$

and

$$z_2(t) = \frac{u_0(\omega)u_1(\bar{\omega})}{\omega^2\bar{\omega}^2}.$$

The function $s(x) = \sin x/x$ and, as before, $\omega = x(1+m^2-2mt)^{1/2}$.

4.2.4 The scalar Perelman approximation

For forward scattering of radiation by a homogeneous sphere the scalar scattering function can be expressed as the following partial wave sum (Sharma *et al.*, 1982; Roy and Sharma, 1996):

$$S(0) = \sum_{l=0}^{\infty}(2l+1)b_l. \quad (4.61)$$

In contrast to the summation from $l=1$ to $l=\infty$ in the event of vector scattering, the summation in scalar formulation is from $l=0$ to $l=\infty$. In addition, $a_l = b_l$ in scalar formulation. Inserting the MPA given by (4.36) into (4.61), a straightforward calculation gives:

$$S(0)_{SPA} = \frac{R^2}{64}\left[2+\frac{4[1-\exp(i\rho)]}{\rho^2}+\frac{4i}{\rho}\exp(i\rho)\right] - \frac{\rho^2}{64}\left[2+\frac{4[1-\exp(-iR)]}{R^2}-\frac{4i}{R}\exp(-iR)\right]. \tag{4.62}$$

The subscript SPA refers to scalar Perelman approximation. The first term on the right-hand side of (4.62) is nothing more than the scattering function for a sphere in the ADA multiplied by a simple m-dependent factor. The second term on the right-hand side of (4.62) can be obtained readily from the first term by replacing ρ by $-R$. Using the extinction theorem, the expression for the extinction efficiency for a nonabsorbing sphere can then be cast in the following form:

$$Q_{SPA}^{ext} = \frac{(m+1)^2}{4}\left[Q_h(\rho) - \frac{\rho^2}{R^2}Q_h(-R)\right]. \tag{4.63}$$

In the limit $x \to \infty$ for a fixed value of ρ the right-hand side of (4.63) is nothing more than the extinction efficiency factor for a sphere in the ADA. Comparison of the expression for extinction efficiency in the SPA given by (4.63) with the expression for extinction efficiency in the MPA given by (4.42) clearly shows that the SPA constitutes considerable simplification over the MPA.

It may be recalled that the RGDA limit is defined as $m \to 1$ for a fixed x. This implies $\rho \to 0$. In this limit (4.62) gives:

$$\text{Re } S(0)_{SPA} = \frac{(m^2-1)^2 x^4}{4}\left[1+\frac{2}{R^2}Q_h(-R)\right], \tag{4.64}$$

and

$$\text{Im } S(0)_{SPA} = \frac{x^3(m^2-1)}{3} \approx \frac{2x^3(m-1)}{3}. \tag{4.65}$$

Comparison of (4.64) and (4.65) with RGDA and MPA amplitudes shows that – whereas the leading term (i.e., Im $S(0)_{SPA}$) agrees with what one obtains in the RGDA and the MPA – the real part of the scattering function agrees with the RGDA only if $R \gg 1$. Thus, in the validity domain of the RGDA ($\rho \ll 1$), the three approximations – namely, the RGDA, the SPA and the MPA – are expected to yield equally good results for scattered intensity because Im $S(0)$ is the leading term. For extinction efficiency, however – which is related to the Re $S(0)$ – the MPA is expected to perform better than the SPA.

A numerical comparison of the MPA and the SPA has been performed by Roy and Sharma (1996) for scattered intensities as well as for extinction efficiency factors. On the basis of this comparison, they proposed a modified SPA (MSPA) in which:

$$S(0)_{MSPA} = S(0)_{SPA} - \phi(m-1), \tag{4.66}$$

and

$$Q_{MSPA}^{ext} = Q_{SPA}^{ext} - \phi(m-1), \tag{4.67}$$

with

$$\phi(m-1) = \tfrac{1}{25}(m-1) + 5(m-1)^2 - 12(m-1)^3 - 2(m-1)^4. \tag{4.68}$$

The MSPA retains the simplicity of scalar approximation while being as accurate as the MPA.

4.2.5 Infinitely long cylinder

For an infinitely long cylinder the PA has been treated only for the perpendicular incidence of light. Here, as in the problem of scattering by a sphere, it is convenient to set forth the scattering coefficients in the following form:

$$b_l = \frac{h_{1l}}{h_{1l} + ih_{3l}}, \qquad (4.69)$$

and

$$a_l = \frac{h_{2l}}{h_{2l} + ih_{4l}}, \qquad (4.70)$$

where

$$\left.\begin{array}{l} h_{1l} = mJ_l'(mx)J_l(x) - J_l(mx)J_l'(x), \\ h_{2l} = J_l'(mx)J_l(x) - mJ_l(mx)J_l'(x), \\ h_{3l} = mJ_l'(mx)N_l(x) - J_l(mx)N_l'(x), \\ h_{4l} = J_l'(mx)N_l(x) - mJ_l(mx)N_l'(x). \end{array}\right\} \qquad (4.71)$$

It is shown in Appendix C that in the limit $m \to 1$ the terms in the denominators of (4.69) and (4.70) can be approximated as follows:

$$h_{1l} = h_{2l} = 0, \qquad (4.72)$$

and

$$h_{3l} = h_{4l} = -\frac{2m^n}{\pi x} = -\frac{2}{\pi x}. \qquad (4.73)$$

With this supposition the summation in the exact series solutions (3.141) and (3.142) can be carried out to yield (Appendix C):

$$T(\theta)_{PA}^{TMWS} = \frac{-i\pi\rho(m+1)J_1(x\omega(\cos\theta))}{4\omega(\cos\theta)}, \qquad (4.74)$$

and

$$T(\theta)_{PA}^{TEWS} = \cos\theta\, T(\theta)_{PA}^{TMWS}, \qquad (4.75)$$

where $\omega(\cos\theta)$ is defined via relation (4.46). For $\theta = 0$, the scattering function for transverse magnetic wave scattering (TMWS) is identical to that for transverse electric wave scattering (TEWS).

The main form of the PA can be obtained by re-writing scattering coefficients (4.69) and (4.70) as,:

$$b_l = \frac{h_{1l}(h_{1l} - ih_{3l})}{h_{1l}^2 + h_{3l}^2}, \qquad (4.76)$$

and

$$a_l = \frac{h_{2l}(h_{2l} - ih_{4l})}{h_{2l}^2 + h_{4l}^2}, \qquad (4.77)$$

and introducing approximation (4.73) into the denominators of (4.76) and (4.77). The resulting expressions for scattering functions involve an infinite sum over products of four Bessel and Hankel functions. Addition theorems involving four Bessel functions of integer order do not seem to exist in the standard literature. In Appendix C we illustrate how this sum over four Bessel (Hankel) functions can be performed to yield the following single-parameter integral representation for scattering functions:

$$T(\theta)_{MPA}^{TMWS} = \frac{\pi x^2 (m^2 - 1)^2}{8} \int_0^{2\pi} d\phi \, \frac{J_1(x\omega(\cos(\theta - \phi))) H_1(x\omega \cos \phi)}{\omega(\cos(\theta - \phi)) \omega \cos(\phi)}, \quad (4.78)$$

and

$$T(\theta)_{MPA}^{TEWS} = \cos\theta \, T(\theta)_{MPA}^{TMWS}, \quad (4.79)$$

where ω is as defined in (4.46) and $H_1 = J_1 - iN_1$. Equations (4.78) and (4.79) constitute the main form of the PA for the scattering of light by an infinitely long homogeneous circular cylinder.

In the Rayleigh limit $x \to 0$, $J_1(x) \to x/2$ and $N_1(x) \to -2/x\pi$, it can easily be verified that in this limit (4.78) and (4.79) give:

$$\text{Re } T(\theta)_{MPA}^{TMWS} = \frac{\pi^2 x^4 (m^2 - 1)^2}{16}, \quad (4.80)$$

and

$$\text{Im } T(\theta)_{MPA}^{TMWS} = \frac{\pi x^2 (m^2 - 1)}{4}, \quad (4.81)$$

respectively. These are, indeed, the expressions for the thin needles in the Rayleigh approximation (van de Hulst, 1957).

For forward scattering $\theta = 0$ and the expression for extinction efficiency in the MPA assumes the form:

$$Q_{MPA}^{ext} = \frac{\pi (m^2 - 1)^2 x}{4} \int_0^{2\pi} d\phi \, \frac{J_1^2(x\omega(\cos \phi))}{\omega^2(\cos \phi)}. \quad (4.82)$$

In writing (4.82), use has been made of the extinction theorem:

$$Q_{ext} = \frac{2}{x} \text{Re } T(0). \quad (4.83)$$

By making a change of variable to $p = x\omega$ one can re-state extinction efficiency in the MPA as:

$$Q_{MPA}^{ext} = \pi (m^2 - 1)^2 x^3 \int_{x|m-1|}^{x|m+1|} dp \, \frac{J_1^2(p)}{p[[p^2 - x^2(m-1)^2][x^2(m+1)^2 - p^2]]^{1/2}}. \quad (4.84)$$

As $m \to 1$ for fixed ρ, the above equation simplifies to:

$$Q_{MPA}^{ext} = 2\pi (m - 1)^2 x^2 \int_{\rho/2}^{\infty} dp \, \frac{J_1^2(p)}{p[p^2 - (\rho/2)^2]^{1/2}}. \quad (4.85)$$

By making the substitution $p^2 - (\rho/2)^2 = s^2$, it may easily be observed that

$Q_{MPA}^{ext} = \pi S_1(\rho)$ – that is, Q_{MPA}^{ext} is nothing more than the well-known expression for the extinction efficiency factor in the ADA.

4.3 HART AND MONTROLL APPROXIMATION

Like the Perelman approximations (PA and MPA), the Hart and Montroll approximation (HMA) (Hart and Montroll, 1951; Montroll and Hart, 1951) also consists in approximating the denominators of the scattering coefficients and, thus, can be applied only to particles with certain specific geometries. The HMA differs from the PA or the MPA in that – while the limit considered in the PA and the MPA is $m \to 1$ – the fundamental assumption underlying the HMA is $l \ll x$.

4.3.1 Homogeneous sphere

When $x \gg l$, the Bessel and Hankel functions may be represented as:

$$j_l(x) \sim \frac{1}{x} \cos\left(x - \frac{(l+1)\pi}{2}\right), \quad (4.86)$$

and

$$h_l^{(1)}(x) \sim (-i)^l \frac{\exp(ix)}{x}, \quad (4.87)$$

where $h_l^{(1)} = j_l(x) - in_l(x)$. Using (4.86) and (4.87) in the denominators of a_l and b_l (equations 4.21 and 4.22), these can be recorded as:

$$h_{1l} + ih_{3l} \sim \frac{i(m+1)m}{2} \exp(-i\rho/2)[1 - (-1)^l \tilde{r} \exp(2imx)], \quad (4.88)$$

and

$$h_{2l} + ih_{4l} \sim \frac{i(m+1)m}{2} \exp(-i\rho/2)[1 + (-1)^l \tilde{r} \exp(2imx)], \quad (4.89)$$

where $\tilde{r} = (m-1)/(m+1)$. It may be observed that in the limit $m \to 1$ (4.88) and (4.89) reduce to:

$$h_{1l} + ih_{3l} \sim h_{2l} + ih_{4l} = im.$$

This description of denominators is then essentially identical to that in the PA. Consequently, approximations (4.88) and (4.89) are valid under both sets of conditions – namely:

$$|m - 1| \ll 1, \quad (4.90)$$

as well as:

$$x \gg l. \quad (4.91)$$

Using the addition theorems given in Appendix B, the infinite series for $S_1(\theta)$ and $S_2(\theta)$ can be summed to yield:

$$S_1(\theta)_{HMA} = \frac{-i\pi(m-1)\exp(i\rho/2)}{m^{1/2}\sin\theta[1 - \tilde{r}^2 \exp(4imx)]}[G_1(\theta) + \tilde{r}\exp(2imx)G_1(\pi - \theta)], \quad (4.92)$$

and

$$S_2(\theta)_{HMA} = \frac{i\pi(m-1)\exp(i\rho/2)}{m^{1/2}\sin\theta[1-\tilde{r}^2\exp(4imx)]}[G_2(\theta) + \tilde{r}\exp(2imx)G_2(\pi-\theta)], \quad (4.93)$$

where G_1 and G_2 are given by equations:

$$\begin{pmatrix} G_1 \\ G_2 \end{pmatrix} = \left(\frac{2m}{\pi}\right)^{1/2} \frac{mJ_{3/2}(x\omega(\cos\theta))\sin\theta}{(x\omega(\cos\theta))^{3/2}} \begin{pmatrix} 1 \\ \cos\theta \end{pmatrix},$$

with $\omega(\cos\theta)$ as defined in (4.46). In the corresponding expression for $G_J(\pi-\theta)$, $\omega(\cos\theta)$ is replaced by $\omega(-\cos\theta)$. Neglecting the terms of relative order $\tilde{r}$ or higher, the extinction efficiency factor in this approximation can be designed as:

$$Q_{HMA}^{ext} = \frac{\pi x^2(m-1)^2}{2m}\left[(m^2+6m^2+1)\Delta_1 - \frac{2(m^2-1)}{x^2}\Delta_2 + x^{-4}\Delta_3\right], \quad (4.94)$$

where

$$\Delta_j = I_j(x(m+1)) - I_j(x(m-1)),$$

and I_1, I_2 and I_3 are:

$$I_1(x) = \frac{1}{2\pi} - \frac{1}{8x}[J_{1/2}^2(2x) + J_{3/2}^2(2x)],$$

$$I_2(x) = \frac{1}{\pi}\left[\frac{\sin 4x}{2x} - \frac{\sin^2 2x}{4x^2} - 1 + c(4x)\right],$$

$$I_3(x) = \frac{1}{\pi}\left[2x^2 + x\sin 4x - \frac{5}{2}\sin^2 2x + c(4x)\right],$$

with

$$c(x) = \int_0^x (1-\cos t)dt/t.$$

The explicit expressions for Δ_1, Δ_2 and Δ_3 then can be re-cast as (Sharma and Somerford, 1996):

$$\Delta_1 = \frac{1}{\pi\rho^2}\left[Q_h(\rho) - \frac{\rho^2}{r^2}Q_h(R)\right],$$

$$\Delta_2 = \frac{1}{\pi}[Q_1(R) + c(R) - Q_1(\rho) - c(\rho)] + \frac{4}{\pi}\left[\frac{\cos R - 1}{R^2} - \frac{\cos\rho - 1}{\rho^2}\right],$$

and

$$\Delta_3 = \frac{2}{\pi}\left[Q_1(R) - Q_1(\rho) + \frac{1}{2}c(R) - \frac{1}{2}c(\rho) + \cos R - \cos\rho\right].$$

It is then straightforward to derive the following relationship:

$$Q_{HMA}^{ext} = \frac{4m}{(m+1)^2}Q_{MPA}^{ext}. \quad (4.95)$$

between the extinction efficiency factors in the HMA and the MPA. The factor $4m/(m+1)^2$ is very close to 1 for $|m-1| \ll 1$. For example, for $m = 1.05$ and

$m = 1.10$ its value is 0.9994 and 0.9975, respectively. The origin of this simple relationship is not transparent immediately. Both approximations consist in summing infinite Mie series with approximate denominators. But, the approximations themselves are apparently quite different. This relationship may be understood in the following way. The HMA holds for $x \gg l$. The main contribution then comes from the rays near the central ray. This has, in fact, led van de Hulst (1957) to suggest that this approximation might be called the "central-incidence approximation". Further, it is also known (see, e.g., van de Hulst, 1957) that the dominant contribution to forward scattering for a large particle when $|m - 1| \ll 1$ arises from near-central rays. This is perhaps the reason for the close relationship between the HMA and the MPA for forward scattering and, hence, for extinction efficiency.

The limiting value of Q_{HMA}^{ext} as $x \to \infty$ can be found to be:

$$Q_{HMA}^{ext}(x \to \infty) = 2 \frac{(m^2 + 1)^2}{(m + 1)^2} \left[1 - \frac{(m^2 - 1)^2}{2m(m^2 + 1)} \log\left(\frac{m + 1}{m - 1}\right) \right]. \quad (4.96)$$

As $m \to 1$, $Q_{HMA}^{ext} \to 2$, as it should be. If one takes the limit $m \to 1$ for a fixed ρ of equation (4.96), the limiting formula is indeed the extinction efficiency obtained in the ADA.

For the case of scalar waves van de Hulst (1957) has cast the $S(\theta)$ in the HMA in the following form:

$$S_{HMA}(\theta) = \frac{im(m - 1)x^3(2\pi)^{1/2}}{1 - \tilde{r}^2 e^{4imx}} \left[e^{-i(m-1)x} \frac{J_{3/2}(v)}{v^{3/2}} - \tilde{r} e^{-i(3m-1)x} \frac{J_{3/2}(\omega)}{\omega^{3/2}} \right], \quad (4.97)$$

and

$$Q_{HMA}^{sca} = \pi x^2 m(m - 1)^2 [\Phi(mx - x) - \Phi(mx + m)], \quad (4.98)$$

where

$$\Phi(u) = \frac{2}{\pi u^2} \left[1 - \frac{2 \sin 2u}{2u} + \frac{2(1 - \cos 2u)}{2u} \right]. \quad (4.99)$$

In calculating scattering efficiency, all terms with $\tilde{r}$ have been neglected. Calculation has been done using equation (3.8).

4.3.2 Infinitely long cylinders

Let us now consider the scattering of light in a normally illuminated infinitely long cylinder in the HMA. When $x \gg l$, one can approximate:

$$J_l(x) \sim \left(\frac{2}{\pi x}\right)^{1/2} \cos\left(x - \frac{l\pi}{2} - \frac{\pi}{4}\right), \quad (4.100)$$

and

$$H_l(x) \sim \left(\frac{2}{\pi x}\right)^{1/2} \exp\left[i\left(x - \frac{l\pi}{2} - \frac{\pi}{4}\right)\right]. \quad (4.101)$$

With this approximation the denominators of b_l and a_l, respectively, become:

110 Other soft-particle approximations [Ch. 4]

$$h_{1l} + ih_{3l} \sim \left(\frac{1}{xy}\right)^{1/2} \frac{i(m+1)}{\pi} \exp\left(\frac{-i\rho}{2}\right)[1 - i\tilde{r}(-1)^l \exp(+2imx)], \quad (4.102)$$

and

$$h_{2l} + ih_{4l} \sim \left(\frac{1}{xy}\right)^{1/2} \frac{i(m+1)}{\pi} \exp\left(\frac{-i\rho}{2}\right)[1 + i\tilde{r}(-1)^l \exp(+2imx)], \quad (4.103)$$

where, as before, $\tilde{r} = (m-1)/(m+1)$. In the limit $|m| \to 1$, (4.102) and (4.103) are identical with the PA. Using the addition theorems:

$$\sum_{l=-\infty}^{\infty} \exp(-il\theta) J_l'(x') J_l(x) = -\frac{J_1(u)}{u}(x' - x\cos\theta), \quad (4.104)$$

and

$$\sum_{l=-\infty}^{\infty} (-1)^l \exp(-il\theta) J_l'(x') J_l(x) = -\frac{J_1(v)}{v}(x' + x\cos\theta), \quad (4.105)$$

the series for the scattering function can be summed (Sharma et al., 1997b) to give:

$$T(\theta)_{HMA}^{TMWS} = -\frac{i\pi x \rho m^{1/2} \exp(i\rho/2)}{2[1 + \tilde{r}^2 \exp(4imx)]} \left(\frac{J_1(x\omega(\cos\theta))}{x\omega(\cos\theta)} + i\tilde{r}\frac{J_1(x\omega(-\cos\theta))}{x\omega(-\cos\theta)} \exp(2imx)\right), \quad (4.106)$$

and

$$T(\theta)_{HMA}^{TEWS} = \cos\theta\, T(\theta)_{HMA}^{TEWS}. \quad (4.107)$$

For $\tilde{r} \ll 1$:

$$T(\theta)_{HMA1}^{TMWS} = \frac{i\pi x \rho}{2} e^{i\rho/2} \frac{J_1(u)}{u}, \quad (4.108)$$

and

$$T(\theta)_{HMA1}^{TEWS} = \cos\theta\, T(\theta)_{HMA1}^{TEWS}. \quad (4.109)$$

But, for the exponential factor in (4.108) the result is identical to that in the PA and the modified RGDA of Gordon (1985). For intensity, even this difference disappears. This, indeed, may be taken as an explanation for the validity of the RGDA outside its validity domain given by the theory.

4.4 EVANS AND FOURNIER APPROXIMATION

The Evans and Fournier approximation (EFA) (Evans and Fournier, 1990, 1994; Fournier and Evans, 1991, 1996) was designed to modify empirically the extinction efficiency factor obtained in the ADA in such a way that it correctly accounts for the behavior of extinction efficiency over the entire range of sizes. This intent has been achieved by using empirical functions obtained through approximation of Mie computation results. Simple expressions for the extinction efficiency for a sphere, a spheroid, randomly oriented spheroids, circular and elliptic cylinders have been obtained in this estimation scheme. The validity of this procedure for the polydispersions of spheres has also been checked.

4.4.1 Homogeneous sphere

The empirical extinction efficiency obtained in the EFA for a homogeneous sphere is:

$$Q_{EFA}^{ext} = Q_R^{ext}\left[1 + \left(\frac{Q_R^{ext}}{TQ_{ADA}^{ext}}\right)^P\right]^{-1/P}, \quad (4.110)$$

where

$$Q_R^{ext} = \frac{24nn'}{F_1(n,n')}x + \left[\frac{4nn'}{15} + \frac{20nn'}{3F_2(n,n')} + \frac{4.8nn'[7(n^2+n'^2)^2 + 4(n^2-n'^2-5)]}{F_1^2(n,n')}\right]x^3$$

$$+ \frac{8}{3}\left[\frac{[(n^2+n'^2)^2 + (n^2-n'^2-2)]^2 - 36n^2n'^2}{F_1^2(n,n')}\right]x^5, \quad (4.111)$$

is the extinction efficiency correct to order x^4 in the Rayleigh approximation ($x \ll 1$, $|mx| \ll 1$). The result (4.111) has been obtained by Penndorf (1960) in which F_1 and F_2 are given by the expressions:

$$F_1(n,n') = (n^2+n'^2)^2 + 4(n^2-n'^2) + 4, \quad (4.112)$$

$$F_2(n,n') = 4(n^2+n'^2)^2 + 12(n^2-n'^2) + 9. \quad (4.113)$$

The parameters P and T are given by the relations:

$$P = A + \frac{\mu}{x}, \quad (4.114)$$

and

$$T = 2 - \exp(-x^{-2/3}). \quad (4.115)$$

When $x \to 0$, P goes to infinity and Q_{EFA}^{ext} equals the Rayleigh formula. As the size parameter increases, Q_R^{ext} becomes very large and Q_{EFA}^{ext} approaches TQ_{ADA}^{ext} – designed to reproduce approximately the large-particle formula of Nussenzveig and Wiscombe (1980). However, if n' is large, Q_R^{ext} outside the Rayleigh region may become negative. In this case one arbitrarily sets the negative coefficient in (4.111) to 0. This ensures the positive growth of Q_R^{ext} as x increases.

The behavior of this approximation in the intermediate region between Rayleigh and ADA limits is controlled by A and μ. By extensive trial and error Evans and Fournier arrived at the following expressions for A and μ:

$$A = \frac{1}{2} + \left[n - 1 - \frac{2}{3}\sqrt{n'} - \frac{n'}{2}\right] + \left[n - 1 + \frac{2}{3}\left(\sqrt{n'} - 5n'\right)\right]^2, \quad (4.116)$$

and

$$\mu = \frac{3}{5} - \frac{3}{4}\sqrt{n-1} + 3(n-1)^4 + \frac{25}{6 + [5(n-1)/n']}. \quad (4.117)$$

Obviously, the above expressions are by no means unique.

4.4.2 Homogeneous spheroids

The extinction efficiency for a spheroidal scatterer of semi-major axis a and semi-minor axis b can be written in the EFA as (Fournier and Evans, 1991):

$$Q^{ext}_{EFA} = Q^{ext}_R \left[1 + \left(\frac{Q^{ext}_R}{TQ_{ADA}} \right)^P \right]^{-1/P}, \qquad (4.118)$$

where

$$P = A + \frac{\mu}{k\tilde{L}}, \qquad (4.119)$$

with $\tilde{L}$ being the distance traveled by the deviated ray through the spheroid. Thus, the phase shift suffered by the central ray is written as:

$$\rho' = k\tilde{L}(m - \cos\vartheta)r,$$

where $r = a/b$ and ϑ is the deflection angle at the boundary. After straightforward but tedious algebra, it can be shown that ρ' can also be put forth as (Fournier and Evans, 1991):

$$\rho' = kb \left\{ \frac{2r}{p} \left[\frac{Ap^2 + Bs}{A^2p^2 + B^2q^2 + 2ABs} \right] \right\} (m - \cos\vartheta), \qquad (4.120)$$

where

$$A = \frac{s^2 + p^2\Delta}{m(p^4 + s^2)}, \qquad B = \frac{s^2(p^2 - \Delta)^2}{m^2(p^4 + s^2)^2},$$

$$\Delta = [m^2(p^4 + s^2) - s^2]^{1/2}, \qquad q = [r^2A^2 + B^2]^{1/2},$$

and

$$s = p^2q^2 - r^2, \qquad p = [A^2 + B^2r^2]^{1/2}.$$

In the limiting cases of $r \to \infty$ (prolate spheroid) it becomes:

$$\rho' = 2kb[(m^2 - \cos^2\Psi)^{1/2} - \sin\Psi],$$

and for $r \to 0$ (oblate spheroid) it becomes:

$$\rho' = 2ka[(m^2 - \sin^2\Psi)^{1/2} - \cos\Psi].$$

The angle Ψ, as before, is the spheroid orientation angle. Numerical comparisons show that these expressions reproduce the exact extinction efficiency factor for randomly oriented spheroids reasonably faithfully. For a single-oriented spheroid, however, predictions are not always reliable.

A general technique for bridging small- and large-particle extinction efficiencies has been developed by Zhao and Hu (2003) for particles with various shapes and sizes. The proposed expression for the extinction efficiency factor is:

$$Q_{ext} = \frac{(Q_{small}^{abs} + Q_{small}^{sca} + c_6(Q_{small}^{sca})^{c_7})Q_{large}^{ext}}{(Q_{small}^{sca} + c_6(Q_{small}^{sca})^{c_7}) + Q_{large}^{ext}} \qquad (4.121)$$

where

$$Q_{small}^{sca} = c_3 \frac{9kV^2}{16\pi P},$$

and

$$Q_{small}^{abs} = c_1 \frac{3kV}{4P} + c_2 \frac{3k^3 V}{4\pi} \exp\left(-c_4 \frac{3k^3 V}{4\pi}\right),$$

with

$$c_1 = \mathrm{Im}\left[\frac{4(m^2 - 1)}{m^2 + 2}\right], \quad c_2 = \mathrm{Im}\left[\frac{4}{15}\left(\frac{m^2 - 1}{m^2 + 1}\right)^2 \frac{m^4 + 27m^2 + 38}{2m^3 + 3}\right],$$

$$c_3 = \frac{8}{3}\left|\frac{m^2 - 1}{m^2 + 1}\right|^2, \quad c_4 = \left|\frac{\mathrm{Im}(m - 1)}{\mathrm{Re}(m - 1)}\right|,$$

and

$$c_5 = c_6 = |m - 1|, \quad c_7 = 2^{c_5},$$

with V as the particle volume. Further:

$$Q_{large}^{ext} = Q_{ADA} \times Z, \qquad (4.122)$$

with

$$Z = 1 + \frac{1}{\frac{2}{Q_{edge}} + [|m - 1|(Q_{ADA} + 1)]}, \qquad (4.123)$$

and

$$Q_{edge} = \frac{c_0}{k^{2/3} P} \int_B R^{1/3} \, ds, \qquad (4.124)$$

as per the prescription of Jones (1957). In (4.124), c_0 is a function of refractive index, in general, and is approximately 0.996 193 for optically soft particles (Nussenzveig and Wiscombe, 1980), R is the radius of curvature of the scattering object profile at the edge that is perpendicular to the incident wave front, ds is the arc-length element along the boundary B and P is the projected area of the scattering object. Numerical comparisons for spheres, spheroids, infinite cylinders and finite cylinders with exact results have found the above formula to be valid for a wide range of sizes in the region $1 < n \leq 2$ and $0 \leq n' \leq 1$. The region of applicability could be as much as $n = 3$ when averaging over orientation or size is performed.

4.5 BOHREN AND NEVITT APPROXIMATION

Whereas the EFA has been designed to construct an expression for the extinction efficiency factor that reproduces reasonably faithfully exact results over the entire domain of size parameter values, the same is achieved for the absorption efficiency of a sphere in an approximation known as the Bohren and Nevitt approximation (BNA). The assumption is that absorption in the sphere is small.

In geometrical optics, the absorption efficiency of a sphere of radius a and relative refractive index m is given by the relation (Bohren and Huffman, 1983):

$$Q_{GO}^{abs} = 2 \int_0^{\pi/2} \frac{T[1 - e^{-\alpha\xi}]}{1 - Re^{-\alpha\xi}} \cos\theta_i \sin\theta_i \, d\theta_i, \tag{4.125}$$

where θ_i is the angle of incidence, T and R are the transmittance and reflectance of unpolarized light obtained from Fresnel formulas, $\xi = 2a\sqrt{n^2 - (\sin^2\theta_i/n)}$ and α is the absorption coefficient of the sphere. It is assumed in (4.125) that the real part of the refractive index is sufficiently large compared with the imaginary part of the refractive index so that the angle of refraction is approximately real. The subscript GO stands for geometrical optics.

Making a change of variable:

$$u = \frac{(n^2 - \sin^2\theta_i)}{n^2},$$

one obtains:

$$Q_{GO}^{abs} = n^2 \int_{\frac{n^2-1}{n^2}}^{1} f(u)[1 - \exp(-2\kappa\sqrt{u})] \, du, \tag{4.126}$$

where $\kappa = 2xn' = 2a\alpha$ and:

$$f(u) = \frac{T(u)}{1 - [1 - T(u)]\exp(-2\kappa\sqrt{u})}. \tag{4.127}$$

It has been noted by Bohren and Nevitt (1983) that if $f(u)$ is set equal to 1 the integral in (4.126) can be evaluated analytically. The errors introduced due to this simplification have been estimated to be at most a few percent. Thus, to a good closeness the absorption efficiency of a sphere is given by:

$$Q_{BNA}^{abs} = C_1 \left[\frac{1}{n^2} - \frac{2}{n^2}\left[e^{-2\kappa\sqrt{n^2-1}/n}\left(1 + \frac{2\kappa\sqrt{n^2-1}}{n}\right) - e^{-2\kappa}(1 + 2\kappa)\right]\right], \tag{4.128}$$

where

$$C_1 = \frac{4n^3}{(n+1)^2 - (n-1)^2 \exp(-2\kappa)}. \tag{4.129}$$

As κ is increased, Q_{BNA}^{abs} approaches the limit $4n/(n+1)^2$, which is the transmittance of a plane surface for normally incident light. The Q_{BNA}^{abs} is obviously not correct in this limit. It is expected to be high by perhaps a few percent (Bohren and Nevitt, 1983).

For a weakly absorbing sphere ($\kappa \ll 1$) the right-hand side of (4.128) can be expanded in powers of κ. The expansion up to third order in κ leads to:

$$Q^{abs}_{BNA} = \tfrac{4}{3}\kappa n^2(1 - \hat{b}^3), \tag{4.130}$$

where

$$\hat{b} = \frac{(n^2-1)^{1/2}}{n}. \tag{4.131}$$

This may be compared with the weak absorption limits of Q^{abs}_{ADA} and Q^{abs}_R (absorption efficiency in the Rayleigh approximation):

$$Q^{abs}_{ADA} = \tfrac{4}{3}\kappa, \tag{4.132}$$

and

$$Q^{abs}_R = \tfrac{4}{3}\kappa \frac{9n}{(n^2+2)^2}. \tag{4.133}$$

Note that – but for an n-dependent multiplicative factor – (4.130), (4.132) and (4.133) are identical. It has been found (Bohren and Nevitt, 1983) that the value of the absorption efficiency factor derived from (4.130) and (4.133) for n in the range 1.0–1.5 does not differ appreciably from unity. The unexpected implication of this is that – subject to restrictions on the refractive index of the sphere – formula (4.130) is valid for geometrical optics as well as the Rayleigh domain. That is, formula (4.128) will yield good results for small as well as large soft particles.

The relationship between (4.130), (4.132) and (4.133) has prompted Flatau (1992) to propose a modified ADA termed the "anomalous diffraction theory ADT". In the ADT, κ occurring in the usual ADA is replaced by a new κ defined as:

$$\kappa_{new} = n^2(1 - \hat{b}^3)\kappa. \tag{4.134}$$

The absorption efficiency factors in this modified ADA have been compared with those obtained by using the standard ADA and the BNA for various κ values by Flatau (1992). The modified ADA was noted to be in excellent agreement with the BNA for all values of κ except near to unity. For κ close to unity, the modified ADA agrees with the standard ADA, which indeed predicts the correct asymptotic behavior of the absorption efficiency factor.

The BNA result (4.128) can also be re-cast in terms of $\mathcal{K}(\omega^*)$ and expressed as (Flatau, 1992):

$$Q^{abs}_{BNA} = C_1[2\mathcal{K}(-2\kappa) - \hat{b}^2\mathcal{K}(-2\hat{b}\kappa)]. \tag{4.135}$$

Flatau (1992) has examined the absorption efficiency for a polydispersion of spherical particles. Regarding gamma size distribution (see Section 3.10), he obtains the following expression for the averaged extinction efficiency:

$$\bar{Q}^{abs}_{BNA} = C_2[2\mathcal{M}(-2\kappa) - \hat{b}^2\mathcal{M}(-2\hat{b}\kappa)]. \tag{4.136}$$

Equations (4.135) and (4.136) suggest that the BNA may be looked upon as a corrected form of the ADA. It reduces absorption efficiency by a factor $\hat{b}^2\mathcal{K}(-2\hat{b}\kappa)$ in (4.135) and by a factor $\hat{b}^2\mathcal{M}(-\hat{2}b\kappa)$ in (4.136) and enhances absorption by a factor C_2/C_1. For $n=1$, $\hat{b}=0$, the correction terms vanish. Functions $\mathcal{K}$ and $\mathcal{M}$ are defined by equations (3.41) and (3,260) respectively.

An approximation that is somewhat similar to the BNA was obtained by Shifrin and Tonna (1992) for weakly refracting small particles. Their approximation reads:

$$Q_{ST}^{abs} = 1 - e^{-(4/3)\kappa n^2(1-\hat{b}^3)}, \tag{4.137}$$

which for small absorption coincides with (4.130).

Simple formulas for the absorption efficiency of a weakly absorbing sphere ($n' < 0.1$) have been obtained by Kokhanovsky and Zege (1995) through the Mie computation results. Relevant expressions have also been quoted in Kokhanovsky and Zege (1997) and in Kokhanovsky (2005):

$$Q_{KZA}^{abs} = \bar{T}\left(1 - \frac{n^2}{8n'^2 x^2}\left[e^{-4n'x\hat{b}}(1+4n'x\hat{b}) - e^{-4n'x}(1+4n'x)\right] - S(n)\left[1 - e^{-4n'x\hat{b}}\right]^2\right), \tag{4.138}$$

where

$$\bar{T} = 1 + (n-1)(1 - e^{-1/(\bar{t}\rho)}),$$

with

$$\bar{t} = [21.2 + 20.1\log n' + 11.1\log^2 n' + \log^3 n']^{-1}, \quad \rho = 2x(n-1),$$

and

$$S(n) = \frac{8n^2(n^4+1)\ln n}{(n^4-1)^2(n^2+1)} - \frac{n^2(n^2-1)^2}{(n^2+1)^3}\ln\frac{(n+1)}{(n-1)} + \frac{3n^7 - 7n^6 - 13n^5 - 9n^4 + 7n^3 - 3n^2 - n - 1}{3(n^4-1)(n^2+1)(n+1)}.$$

An alternative expression – obtained directly within the complex angular momentum theory for weakly absorbing soft particles ($n \leq 1.2$) (Kokhanovsky, 1995) – is:

$$Q_K^{abs} = Q_{BNA}^{abs} + Q_{edge}^{abs}, \tag{4.139}$$

where

$$Q_{edge}^{abs} = 4nn'x\left[\cos^{-1}\left(\frac{1}{n}\right) - \frac{1}{n^2}\sqrt{n^2-1}\right]. \tag{4.140}$$

For particle size distribution $f(a)$, this gives:

$$\sigma_{abs} = \frac{4\pi n'}{\lambda}A(n)C_v, \tag{4.141}$$

where C_v is the volumetric concentration defined as:

$$C_v = \frac{4\pi N}{3}\int_0^\infty f(a)a^3\,da, \tag{4.142}$$

and $A(n)$ is,:

$$A(n) = n^2(1-\hat{b}^3) + \frac{3}{2}\left(n\cos^{-1}\frac{1}{n} - \hat{b}\right). \tag{4.143}$$

As $n \to 1$, $A(n) \to 1$, equation (4.141) gives $\sigma_{abs} = 4\pi n' C_v/\lambda$, which is simply the absorption efficiency in the RGDA (Kokhanovsky and Zege, 1997).

4.6 NUSSENZVEIG AND WISCOMBE APPROXIMATION

An approximate formula which has been essentially designed for large spherical particles is due to Nussenzveig and Wiscombe (1980) and is called the "Nussenzweig and Wiscombe approximation" (NWA). Since this is an asymptotic approximation in size parameter, it is categorically different from the class of relations being discussed herein – namely, soft-particle approximations – which are basically an outcome of the inequality $|m - 1| \ll 1$. However, since the NWA does not preclude the condition $|m - 1| \ll 1$, one may be legitimately interested to know how this formula compares with other soft-particle relations. The expression for the extinction efficiency in this approximation is:

$$Q_{NWA}^{ext} = 2 + 1.992\,3861 x^{-2/3} - 0.715\,3537 x^{-4/3}$$

$$- 0.332\,0643 \left(\frac{3^{1/2}}{2}\right) \frac{(m^2 + 1)(2m^4 - 6m^2 + 3)}{(m^2 - 1)^{3/2}} x^{-5/3} - \frac{16m^2 \sin \rho}{(m+1)^2 \rho}$$

$$- \frac{16m(3m - m^2 - 1)\cos \rho}{\rho^2} - \frac{8}{x}\sum_{j=1}^{\infty} \frac{(m-1)}{(2j + 1 - m)}$$

$$\times \left[\frac{m-1}{m+1}\right]^{2j} \sin 2(m_1 + 2jm)x + O(1/x^2) + \text{ripple}. \quad (4.144)$$

Using the optical theorem $Q_{ext} = (4/x^2)\,\text{Re}\,S(0)$ and the fact that $S(0, -k) = [S(0, k)]^*$ one can obtain Im $S(0)$ as (Attard et al., 1986):

$$\frac{4}{x^2}\,\text{Im}\,S(0) = -3^{1/2}(1.992\,3861 x^{-2/3} - 0.715\,3537 x^{-4/3}) + \frac{2(m^2 + 1)}{x(m^2 - 1)^{1/2}}$$

$$- 0.332\,0643 \left(\frac{3^{1/2}}{2}\right) \frac{(m^2 + 1)(2m^4 - 6m^2 + 3)}{(m^2 - 1)^{3/2}} x^{-5/3} - \frac{16m^2 \cos \rho}{(m+1)^2 \rho}$$

$$- \frac{16m(3m - m^2 - 1)\cos \rho}{\rho^2} - \frac{8}{x}\sum_{j=1}^{\infty} \frac{(m-1)}{(2j + 1 - m)}$$

$$\times \left[\frac{m-1}{m+1}\right]^{2j} \sin 2(m_1 + 2jm)x + O(1/x^2) + \text{ripple}. \quad (4.145)$$

It may be mentioned here that the original paper of Nussenzveig and Wiscombe (1980) contained a misprint which was subsequently corrected by Nussenzveig (1984). The corrected form is also contained in the work of Attard et al. (1986) and Roy and Sharma (1996).

4.7 PENNDORF–SHIFRIN–PUNINA APPROXIMATION

Let us begin by recalling the following definitions for the scattering of light by a homogeneous sphere:

$$i(0) = [\text{Re } S(0)]^2 + [\text{Im } S(0)]^2.$$

Since:

$$\text{Re } S(0) = \frac{x^2}{4} Q_{ext},$$

Penndorf (1962) has proposed the approximation:

$$i_p(0) = \frac{x^4}{4}\left(\frac{Q_{ext}}{2}\right)^2, \quad (4.146)$$

if $\text{Re } S(0) \gg \text{Im } S(0)$. Further, since $i_f(0) = x^4/4$ in the Fraunhofer diffraction approximation, equation (4.146) may be written as:

$$i_p(0) = i_f(0)\left(\frac{Q_{ext}}{2}\right)^2. \quad (4.147)$$

On the basis of the above result Shifrin and Punina (1968) have proposed the following formula for non-forward scattering:

$$i_{pspa}(\theta) = i_f(\theta)\left(\frac{Q_{ext}}{2}\right)^2. \quad (4.148)$$

The factor $(Q_{ext})^2/4$, therefore, may be looked upon as the factor that scales the Fraunhofer diffraction curve to those predicted by Mie solutions. From a detailed analysis of the errors in the $i(0)_{pspa}$, Fymat and Mease (1981) have derived the following semi-empirical formula:

$$i_{FMA}(\theta) = I_{pspa}(\theta) f_1 f_2, \quad (4.149)$$

where

$$f_1^{-1} = f^{-1}[(m-1)x] = 1 - J_0^2[(m-1)x],$$

for $(m-1)x \leq a$ or $b \leq (m-1) \leq c$ or 1 otherwise and:

$$f_2^{-1} = f^{-1}[2(m-1)x] = 1 - J_0^2[2(m-1)x],$$

for $2(m-1)x \leq d$ and 1 otherwise. Here a, b, c, d are refractive index dependent constants. For $m = 1.33$, these are found to be $a = 3.63$, $b = 5.52$, $c = 6.6$ and $d = 2.40$. The formula has been found to be reasonably accurate for near-forward scattering.

4.8 NUMERICAL COMPARISONS

The range of validity of the RGDA for a homogeneous sphere has been investigated by many workers. Mention may be made of papers by Farone et al. (1963), Heller (1963) and Kerker et al. (1963). A detailed investigation by Farone and Robinson

(1968) reveals that for $m \leq 1.25$ the approximate formula for the extinction efficiency factor in the RGDA is better than 10% for $x < 1.0$. For a fixed value of m, accuracy decreases as x increases. Table 4.1 shows the percent error in forward scattered intensities. For $\rho \ll 1$, errors are reasonably small. The angular variation of the intensity is also produced quite accurately in the RGDA. In particular, reproduction of the positions of the first few extrema is very close to exact positions. It is well-known that the positions of the extrema may be used to determine the size of the scatterer. The errors introduced by the use of the RGDA in assessing the size of the scatterer from the position of extrema are discussed in detail in Chapter 5, where applications of the EA are described. The errors in the RGDA have also been examined for a homogeneous, infinitely long cylinder at perpendicular incidence. The percent errors in forward scattered intensity are shown in Table 4.2. Errors in the corresponding formulas for a spheroid have been examined by Barber and Wang (1978).

Perelman (1991) has checked the closeness of Q_{MPA}^{ext} to exact results over a large range of m and x values for a homogeneous sphere. The outcome is exhibited in Table 4.3. The table specifies maximum ρ values when error is less than 5% for a given m value. For $m \leq 1.06$ errors are less than 5% for all values of x, except possibly at points between $0 < x < 2$. In this region, error can be between 5 and 25%. Also, it has been noted that (4.42) and (4.57) are on the whole of the same accuracy (Perelman, 1978). The errors in Q_{MPA}^{ext} for scattering of normally incident light on an infinitely long homogeneous cylinder have also been examined. For $m \leq 1.05$ and $x > 2.0$, maximum error is less than 2.27% (Sharma et al., 1997b). Figure 4.1 depicts these errors for various values of the relative refractive index. Granovskii and Ston (1994a)

Table 4.1. The percent errors in $i(0)_{RGDA}$ for scattering of unpolarized incident radiation by a homogeneous sphere of $m = 1.05$.

x	0.2	0.4	0.6	0.8	1.0	5.0	10.0	20.0	25.0
Error	0.14	0.87	1.72	2.72	3.54	7.61	4.38	−14.61	−31.03

Table 4.2. The percent errors in $i(0)_{RGDA}$ for $m = 1.05$ for a homogeneous, infinitely long cylinder at perpendicular incidence.

x	0.2	0.4	0.6	0.8	1.0	5.0	10.0	20.0	25.0
TMWS	0.79	1.91	2.60	2.59	1.99	−1.75	−7.43	−33.48	−57.88
TEWS	0.43	1.12	1.74	2.19	2.91	6.32	1.92	−21.38	−43.49

Table 4.3. The ρ values below which the extinction in the MPA gives errors that are <5%.

m	1.00–1.06	1.08	1.10	1.12	1.14	1.16	1.18	1.20	1.22
$\rho(m)$	∞	128	60	31	25	23	18	15	14

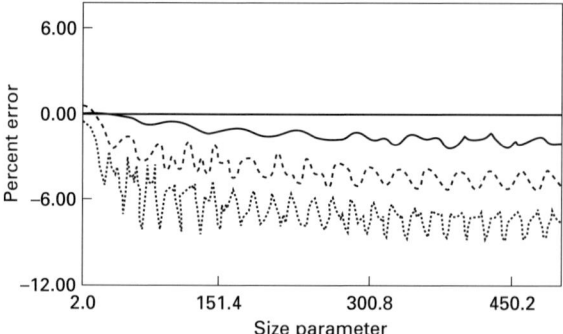

Figure 4.1. Percent error in extinction efficiency *versus* x for MPA for $m = 1.05$ (solid curve), $m = 1.10$ (broken curve) and $m = 1.15$ (dotted curve).
From Sharma *et al.*, 1997b.

have compared the extinction efficiency results of their integral representation of the MPA with those of Mie theory, the RGDA and ADA. They considered ρ values up to 10 and $m = 1.1$. Their predictions are within 2% of exact Mie theory.

Comparison of the IPA with the ADA and Mie theory for extinction efficiencies show that, for very small size parameters, errors are larger than those when using the RGDA. For large phase shifts, the IPA reproduces very well both the position and the height of the maxima and minima of the extinction curve. In comparison the extinction efficiency factor predicted by the ADA is smaller than that given by exact theory. It is found that the IPA may be used for arbitrarily large (infinite) values of x for $m < 1.03$ (error is $<1\%$), for $m < 1.10$ (error is $<5\%$) and $m < 1.22$ (error is $<10\%$) (Perelman and Voshchinnikov, 2002). The dependence of x on m has been estimated as $x \leq 240m^{-15.6}$ if error is to be less than 1%, as $x \leq 160m^{-9.30}$ if error is to be within 5% and as $x \leq 270m^{-9.99}$ if error is to be within 10%.

A typical example of variation of angular scattered intensity in the MPA is depicted in Figure 4.2. In this figure $x = 20.0$ and $m = 1.50$. It can be seen that

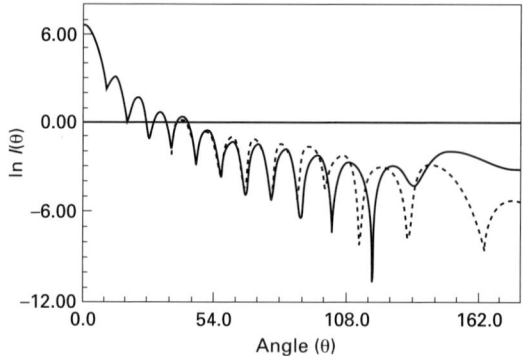

Figure 4.2. Comparison of variation of $ln[I(\theta)]$ with θ in the MPA (broken curve) with the corresponding exact result (solid curve) for $m = 1.05$ and $x = 20.0$.
From Sharma *et al.*, 1997b.

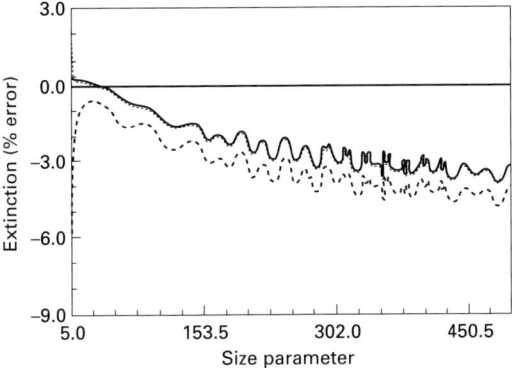

Figure 4.3. Percent error in extinction efficiencies against size parameter for $m = 1.06$. Solid line: MPA, broken line: SPA, dotted line: MSPA.
From Roy and Sharma (1996).

near-forward scattering is reproduced very well. The positions of maxima and minima as well as their amplitudes are reproduced quite accurately for the first few extrema.

An illustrative comparison of percent errors in the MPA, the SPA and the MSPA for Q_{ext} against size parameter x is shown in Figure 4.3 for $m = 1.06$. The accuracy of the MPA and the MSPA is nearly identical. Both approximations are more accurate than the SPA. The same trend has been noted for percent errors in forward scattered intensity also. The magnitude of errors in the MPA and the MPSA is marginally higher for scattered intensity. As expected, errors increase as m increases.

The extinction efficiency in the MPA is compared with the ADA and the FCII in Table 4.4 for $m = 1.06$. The errors in the MADA and the MPA increase with increasing x tending to a constant value. Shortcomings in the ADA decrease with increasing x, approaching 0 as $\rho \to \infty$. The ρ value above which the ADA is the most

Table 4.4. Percent error in various approximate methods in $1.0 \leq x \leq 10.0$ for Q_{ext} for a homogeneous sphere. The relative refractive index is $m = 1.06$.

x	SPA	MPA	ADA	FCII
1.0	−60.83	−7.83	−142.21	−155.73
2.0	−42.30	−3.79	−40.93	−48.79
3.0	−19.92	−0.35	−16.20	−22.69
4.0	−12.53	−0.09	−7.83	−13.85
5.0	−8.54	0.05	−3.21	−8.97
6.0	−6.03	0.31	−0.64	−6.26
7.0	−4.66	0.28	0.83	−4.70
8.0	−3.70	0.26	1.89	−3.58
9.0	−2.94	0.32	2.65	−2.77
10.0	−2.46	0.30	3.17	−2.23

From Roy and Sharma (1996).

Table 4.5. Percent error in various approximate methods in $1.0 \leq x \leq 10.0$ for $i(0)$ for a homogeneous sphere. The relative refractive index is $m = 1.06$.

x	SPA	MPA	ADA	FCII
1.0	−10.29	−21.03	2.25	−8.96
2.0	−6.90	−14.20	5.08	−5.01
3.0	−4.18	−8.91	7.24	−3.39
4.0	−3.39	−6.81	8.08	−2.46
5.0	−2.86	−5.50	8.54	−1.95
6.0	−2.38	−4.46	8.92	−1.53
7.0	−2.19	−3.86	9.13	−1.29
8.0	−2.04	−3.40	9.26	−1.15
9.0	−1.87	−2.97	9.39	−1.00
10.0	−1.78	−2.68	9.49	−0.89

From Roy and Sharma (1996).

accurate approximation have been delineated by Perelman (1978). For a given relative error ϵ, approximation (4.57) is preferable in the domain $0 < x < x(m, \epsilon)$. Generally speaking, the function $x(m, \epsilon)$ decreases as $|m - 1|$ increases. The rate of decrease of $x(m, \epsilon)$ is not uniform. For example, $x(1.02, 0.002) > 100$, $x(1.10, 0.01) = 55$, $x(1.10, 0.03) = 90$, $x(1.20, 0.07) = 85$ and so on. A typical comparison for percent errors in forward scattered intensities is shown in Table 4.5.

For x close to 1, the percent errors in various approximations for extinction efficiency are shown in Table 4.4. These can be noted to be quite large. No approximate formula performs well except MPA. This is expected because only the Q_{MPA}^{ext} reduces to Q_{RGDA}^{ext} for $x \leq 1$. For scattered intensity, the leading contribution in the limit $x \to 0$ arises from the imaginary part of the scattering function and this is reproduced correctly in all approximations.

The numerical computations of Q_{MPA}^{ext} show that it overestimates the true extinction efficiency factor. Generally, the same is true for Q_{ext}^{HMA}. But, as $4m/(m + 1)^2$ is less than 1, Q_{ext}^{HMA} is expected to give slightly better estimates. Indeed, the error in Q_{ext}^{HMA} for a homogeneous sphere is less than 5% for any x as long as $m \leq 1.10$. This may be contrasted with Q_{MPA}^{ext} where the corresponding upper limit is $m = 1.06$. The errors in scattered intensities in the HMA and the HMA1 have also been examined for an infinitely long cylinder at perpendicular incidence (Sharma, 1994). The range of m and x values considered were $1.0 \leq x \leq 50$ and $1.05 \leq m \leq 1.50$. The HMA and the HMA1 agree with exact results very well for $\theta \leq 60°$ except at the positions of minima. Further, the only difference between the HMA and the HMA1 appears to be the fact that the minima appear to be slightly deeper in the HMA1; otherwise, no significant difference from the HMA has been noted. Table 4.6 exhibits a characteristic comparison of percent errors in the forward scattered intensities in the HMA1 with some other approximations for $m = 1.05$.

Figure 4.4 shows a contour plot of maximum percent error in Q_{EFA}^{ext} (reproduced from Evans and Fournier, 1991) for a homogeneous sphere. The range of the real part

Table 4.6. Percent error in various approximations in $1.0 \leq x \leq 25.0$ for $i(0)$ for a homogeneous sphere. The relative refractive index is $m = 1.05$.

x	HMA1	EA	ADA	FCEA
1.0	2.10	2.05	6.77	2.05
3.0	0.06	0.10	4.88	0.05
5.0	−0.10	0.11	4.83	0.01
10.0	−0.82	0.22	4.69	−0.17
15.0	−1.74	0.58	4.59	−0.32
20.0	−3.33	1.25	4.59	−0.37
15.0	−5.32	2.12	4.58	−0.46

From Sharma (1994).

of refractive index is from 1.0 to 2.0 and the imaginary part of the refractive index ranges from 10^{-6} to 10. The agreement with exact results can be seen to be very good. For example, for $n \leq 1.62$ the relative error in the EFA does not exceed 20% (Kokhanovsky and Zege, 1997). For a polydispersion of scatterers the errors decrease considerably. For the same domain of m values the errors for the typical particle size distribution encountered in many atmospheric problems is less than 3%. Numerical comparisons have been performed for Q_{EFA}^{ext} for randomly oriented spheroids as well. The percent error in Q_{EFA}^{ext} was studied by Fournier and Evans (1991) for the semi-major axis parameter a, varying from 0.1 to 30. The refractive index range was taken to be the same as that in Figure 4.4 and the aspect ratio r was taken to be 2. For $r = 2$, the worst case error is $\sim$30% ($n < 1.02$). Error is considerably improved (<10%) when either n is larger or k is increased.

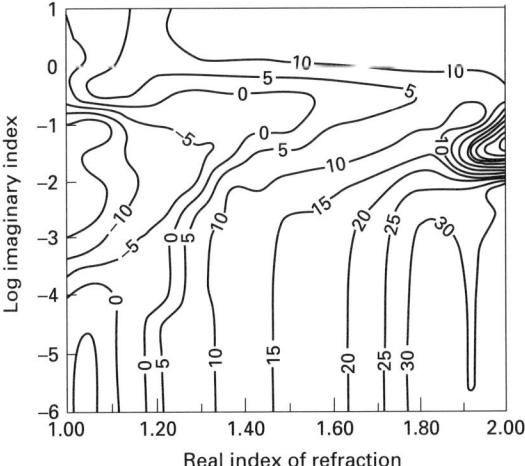

Figure 4.4. Contour plot over the complex index of the refraction plane for extinction efficiency. Maximum percent error between Mie theory and the EFA.
From Evans and Fournier (1990).

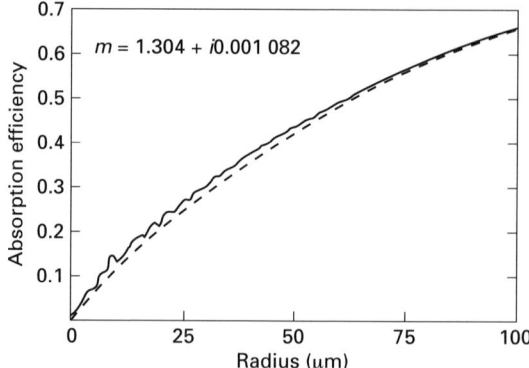

Figure 4.5. Absorption efficiencies of water droplets at $\lambda = 2.0\,\mu$m calculated exactly (solid line) and approximately (dashed line).
From Bohren and Nevitt (1983).

Figure 4.5 shows a comparison of absorption efficiency from (4.128) with exact results for a water droplet of $m = 1.304 + i0.001\,082$. The size range examined was $a = 0$–$100\,\mu$m at $\lambda = 2\,\mu$m. Clearly, the BNA is adequate over the entire size range. Flatau (1992) has compared Q^{abs}_{BNA}, Q^{abs}_{ADA} and the corrected Q^{abs}_{ADA} (equation 4.135) as a function of κ for the fixed real part of the refractive index for a homogeneous sphere. The ADA differs from the BNA in most of the κ region. The corrected ADA, however, gives excellent agreement with the BNA – up to $\kappa \sim 0.5$.

Roy and Sharma (1996) have numerically compared the extinction efficiency and forward scattered intensity in the NWA with exact results, as well as with the PA, the MPA, the SPA and the MSPA for nonabsorbing spheres. The last two terms on the right-hand side of (4.144) and (4.145) have been ignored in numerical calculations. It has been observed that:

(i) If x is greater than a certain value (say, $x_>$), the NWA is better than any of the other approximations considered.
(ii) The value of $x_>$ is dependent on $|n-1|$. It decreases as $|n-1|$ increases. For example, we observe that for $m = 1.02$, $x_> = 374$; $m = 1.06$, $x_> = 82$ and for $m = 1.16$, $x_> = 26$.
(iii) The extinction efficiency becomes less than 0 in the NWA for x values less than a certain value (say $x_<$). But, then, the NWA formula is not expected to be valid in this region of small x values. This value $x_<$ decreases as $|n-1|$ increases. For example, when $m = 1.02$, $x_< = 42$; $m = 1.06$, $x_< = 14$ and for $m = 1.16$, $x_< = 5$.

Attard et al. (1986) have investigated the accuracy of the real and imaginary parts of $S(0)_{NWA}$. The series of refractive indexes examined consisted of real parts 1.1, 1.33, 1.5 and 2.0, and imaginary parts of 0.0, 0.01, 0.1 and 0.5. The conclusion from this study is that the absolute accuracy of the Im $S(0)_{NWA}$ is comparable with that of the accuracy of Re $S(0)_{NWA}$, which has been thoroughly investigated by Nussenzveig

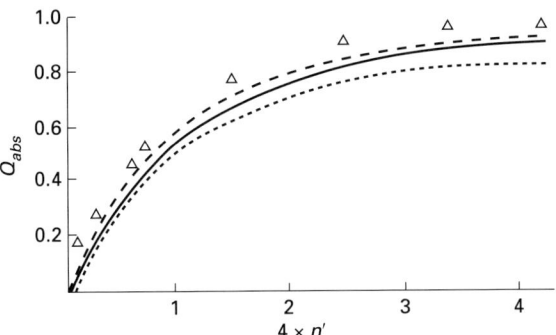

Figure 4.6. Comparison of ZKA (solid line), BNA (dotted line), SA (dashed line) and Δ (Mie calculations).
From Kokhanovsky and Zege (1997).

and Wiscombe (1980) and by Roy and Sharma (1996). But, because $\mathrm{Im}\, S(0)_{NWA}$ is significantly smaller than $\mathrm{Re}\, S(0)_{NWA}$, fractional error in the $\mathrm{Im}\, S(0)_{NWA}$ is much larger than that in the $\mathrm{Re}\, S(0)_{NWA}$.

The error in Q_{KZA}^{abs} given by (4.138) does not exceed 10% for $x \geq 10.0$ for typical aerosol refractive indexes $n = 1.2$–1.55. Kokhanovsky and Zege (1997) have also compared the absorption efficiencies obtained in their own approximation – the Zege and Kokhanovsky (1988) (ZKA) – the Shifrin approximation (SA) and the BNA with exact Mie scattering results. The parameters in these calculations were $n = 1.34$, $x = 20$–200, $n' = 10^{-3}$ for $2n'x < 1$ and $n' = 10^{-2}$ for $2n'x > 1$. The SA was found to be the least erroneous by Kokhanovsky and Zege (1997) (see Figure 4.6). They have also evaluated the accuracy of relation (4.141). Errors are less than 6% at $n < 1.2$, $\kappa < 10^{-4}$ at $\lambda = 0.55\,\mu\mathrm{m}$ and $4\pi x_{eff}/\lambda = 1$–$10\,\mu\mathrm{m}$, where $x_{eff} = 2\pi a_{eff}/\lambda$ with a_{eff} defined as $a_{eff} = \langle a^3 \rangle / \langle a^2 \rangle$ is the Sauter mean radius (Bayvel and Jones, 1981). Further, $\langle a^n \rangle$ is defined as:

$$\langle a^n \rangle = \int a^n f(a)\, da,$$

is the nth moment of the distribution.

Shifrin and Punina (1968) have examined the PSPA numerically as a function of x. For transparent water aerosols and for $x > 5$ the approximation agrees well with exact results. In the domain $1.0 < x < 5.0$, however, results show a large error. In this range the FMA shows a dramatic improvement over the PSPA in the forward direction.

5

Applications of eikonal-type approximations

Soft-particle approximations have found widespread applications for particle characterization and in other studies relating to electromagnetic wave propagation in an ensemble of particles. This is particularly true for the anomalous diffraction approximation (ADA) and the Rayleigh–Gans–Debye approximation (RGDA). Further, since the RGDA and its applications are available in the literature in detail in a number of books and articles, we restrict ourselves to minimal reference to this formulation. Thus, in this chapter we treat some applications of the anomalous diffraction/eikonal-type approximations with respect to problems of scattering and absorption of radiation by soft particles. Other soft-particle approximations have not attracted much attention as far as applications are concerned.

5.1 PARTICLE SIZE DETERMINATION

5.1.1 One particle at a time

Over a period of time a number of experimental methods have been developed that employ the "one particle at a time" technique for particle size determination. A well-known arrangement of this type utilized for the analysis of particles is the flow cytometer. In a flow cytometer, one particle at a time passes through the scattering region and the scattering pattern is analyzed to typecast the particle. Up to 300,000 particles per minute can be analyzed in a flow cytometer (Shepelevich *et al.*, 1999).

For analysis of the measured scattering pattern, suitable approximate methods are required. From our earlier discussions of the validity domain of the eikonal approximation (EA) and its modified variants in Chapter 3, it is clear that for intermediate-size soft particles it is eikonal-type approximations that lead to best predictions for the small-angle scattering pattern. For large particles the Fraunhofer diffraction prescription can be employed more conveniently – particularly, for absorbing particles. For small-size particles, the Rayleigh approximation or the RGDA would turn out to be more simple.

Comparison of Fraunhofer diffraction and the exact solution for the scattered intensity of an infinitely long cylindrical particle of circular cross-section has been performed by Powers and Somerford (1979) for fibre size determination based on the positions of the maxima or minima. The Fraunhofer approximation is found to agree well with the exact theory for absorbing cylinders of $x \geq 20$, whilst for transparent fibres having $x \leq 100$ the agreement is poor. However, the apparent diameter of the transparent fibre is still a useful quantity since the true diameter may be obtained by applying a suitable correction factor (Powers and Somerford, 1982).

For particles with $x \leq 20$, an obvious choice is to employ the EA or one of its modified forms. But, since this is a small-angle approximation, a signature of the scattering pattern is needed for particle size determination that depends on small scattering angles. Two such signatures are (*i*) the intensity ratio at two angles in the forward lobe – that is, the slope of the forward lobe – and (*ii*) the positions of minima and maxima in the scattering pattern.

Intensity ratio technique

Hodkinson (1966) has noted that measurement of the ratio of scattered intensities at a pair of convenient angles within the forward lobe can give a useful estimate of the size of the scattering particle. Theoretical errors in spherical particle sizing employing this technique and the Fraunhofer diffraction formula were quantified by Boron and Waldie (1978). Following this work, Sharma and Somerford (1983a) examined the variation of $R(\theta_1, \theta_2) = i(\theta_1)/i(\theta_2)$ with x for the EA, the first-order corrected eikonal approximation (FCI) and the RGDA against the exact $R(\theta_1, \theta_2)$ for transverse magnetic wave scattering (TMWS) at selected angle pairs. The EA and the FCI were found to be in close agreement with exact graphs. A typical comparison is shown in Figure 5.1. Clearly, accurate fibre size determination should be possible from a measurement of $R(\theta_1, \theta_2)$.

Measurement of $R(\theta_1, \theta_2)$ for a known m allows the unknown size parameter to be determined from a theoretical graph of the type shown in Figure 5.1. The errors involved in using the FCI in place of exact theory for fibre size determination have been assessed by Sharma and Somerford (1983a). Percent error as a function of size parameter for $m = 1.05$ and $m = 1.15$ for angle pairs $(5°, 2.5°)$, $(10°, 5°)$ and $(20°, 10°)$ are shown in Tables 5.1 and 5.2, respectively. The errors are indeed small. Similar estimates have been made for spherical particles too (Sharma *et al.*, 1984a, b) and are also given in Tables 5.1 and 5.2. The errors have been calculated in the following way. For a given size parameter, $R(\theta_1, \theta_2)$ values at two appropriate angles in the forward lobe were computed using the exact theory. Any divergence between the exact theory and FCI predictions was obtained by applying the intensity ratio computed by the exact theory to the FCI graph of the type shown in Figure 5.1. For the accurate fiber size determination for $x < 20.0$ and $m < 1.15$ a suitable angle pair was noted to be $(5°, 2.5°)$. The percent error in x determination has been defined as $(x_{ex} - x_{FCI}) \times 100/x_{ex}$. The maximum error for fibre size determination in the range $3 \leq x \leq 25$ for $m < 1.05$ for FCI is about 0.4%, which is indeed very small.

Sec. 5.1] **Particle size determination** 129

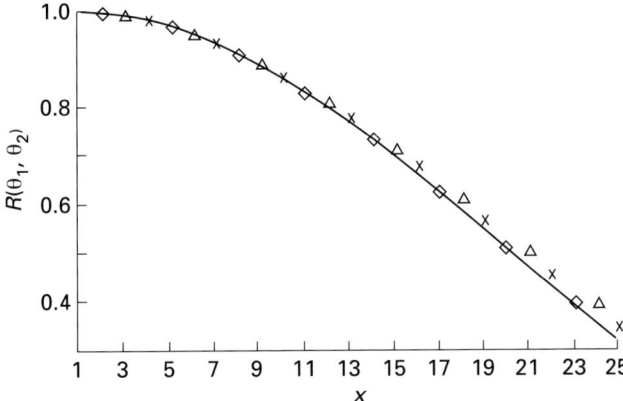

Figure 5.1. Scattered intensity ratio $R(\theta_1, \theta_2)$ *versus* size parameter for an infinitely long homogeneous cylinder of $m = 1.05$ and for the angle pair $(5°, 2.5°)$. Solid line: exact theory; triangles: RGA; crosses: EA; diamonds: FCI.
From Sharma and Somerford (1983a).

Table 5.1. Percent error in size parameter determination for $m = 1.05$ in using FCI.

			$(\theta_1 - \theta_2)°$		
		Infinite cylinder		*Homogeneous sphere*	
x	2.5–5	5–10	10–20	2.5–5	5–10
1.0	2.05	1.65	0.21	−81.60	−82.10
3.0	0.38	0.01	−1.38	−10.40	−10.30
5.0	0.27	−0.13	−1.60	−3.36	−3.37
7.0	0.19	−0.17	−1.71	−1.39	−1.28
9.0	0.17	−0.23	−1.91	−0.60	−0.51
11.0	0.15	−0.24	−3.41	−0.23	−0.15
13.0	0.14	−0.25		−0.25	−0.05
15.0	0.12	−0.15		−0.09	0.17
17.0	0.12	−0.30		−0.16	0.25
19.0	0.10	−0.76		−0.21	0.34
21.0	0.08			0.24	0.39
23.0	0.20			0.26	0.65
25.0	0.10			0.30	

From Sharma and Somerford (1983a) and Sharma *et al.* (1984a).

Table 5.2. Percent error in size parameter determination for $m = 1.15$ in using FCI.

| | \multicolumn{3}{c}{$(\theta_1 - \theta_2)°$} | | |
| | Infinite cylinder | | | Homogeneous sphere | |
x	2.5–5	5–10	10–20	0–5	5–10
1.0	5.72	5.42	4.45	−73.2	−73.1
3.0	1.52	1.07	−0.53	−6.15	−6.02
5.0	1.26	0.86	−0.67	−0.65	−0.53
7.0	0.97	0.57	−1.04	1.18	0.48
9.0	1.30	0.40		0.95	1.04
11.0	0.52	0.25		1.00	1.10
13.0	0.23	−0.66		0.93	1.03
15.0	−0.21			0.80	0.78
17.0	−0.67			0.70	
19.0	−1.23			0.65	
21.0	−1.80			0.76	
23.0				1.21	
25.0				3.79	

From Sharma and Somerford (1983a) and Sharma et al. (1984a).

The method does not apply to small as well as large values of x. For smaller x values the slope at forward angles is extremely small and the error in using a graphical technique becomes significant. For larger particles, the forward lobe shrinks to very small angles. The choice of θ_1 and θ_2 is then also restricted to small angles. But, at small angles, errors in practical measurements of the scattered intensity may become large. The situation may be improved to some extent by using radiation of longer wavelength. Alternatively, one may also choose to use an index matching liquid to bring the relative refractive index of the fibre closer to unity. It should be noted that the position of first minimum increases with decreasing relative refractive index.

Formulas relating to particle size and extrema positions
It is well-known that knowledge of the positions of maxima and minima in a light scattering pattern can be used to determine scatterer size. To this end what is required is a theory that relates size with positions of maxima and minima in a simple way. For soft particles the relations have been obtained based on eikonal-type approximations and Rayleigh–Gans–Debye-type approximations.

Chen (1994) has deduced the relations between the positions of minima in the scattering pattern and the size of a nonabsorbing homogeneous dielectric spherical scatterer employing the generalized EA. Since the EA and its variants are small-angle approximations, only use of the positions of the first few minima will yield accurate results. Starting from the scattering function (3.85), one can write the scattered intensity and examine the positions of extrema in the scattering pattern. This leads to a formula that gives the positions where the scattered intensity becomes the minimum. The locations of the first two minima in the scattering pattern are given by the following relations:

$$2a \sin \theta_1 = \left[1.220 - \frac{(m-1)/(m+1) + 15.87 \sin y_1/y_1}{\pi(1.667 - 12.43 \sin y_1/y_1)}\right] \lambda, \quad (5.1)$$

and

$$2a \sin \theta_2 = \left[2.333 - \frac{(m-1)/(m+1) - 50.0 \sin y_2/y_2}{\pi(2.15 + 21.50 \sin y_2/y_2)}\right] \lambda, \quad (5.2)$$

where

$$y_1 = 2x\sqrt{m^2 + 1 - 2m\sqrt{1 - (1.916/x)^2}},$$

and

$$y_2 = 2x\sqrt{m^2 + 1 - 2m\sqrt{1 - (3.508/x)^2}}.$$

The predictions of (5.1) and (5.2) are found to be in excellent agreement with exact results – especially, for first minimum and for size parameters above about 50. Clearly, equations (5.1) and (5.2) provide good prescriptions for obtaining information about scatterer sizes from the position of minima in the scattering pattern.

It is known that the positions of the minima do not uniquely determine the size of the scatterer even for a given refractive index. That is, for a given m, it is possible to have a minimum at the same angle for different size parameters. This difficulty may be overcome by determining an x which simultaneously satisfies (5.1) and (5.2). In case further discrimination is needed, help of a similar equation for the third minimum may be taken.

A formula analogous to (5.1) but based on the EA has been obtained by Sharma et al. (1997a). It is given by the relation:

$$x \sin \theta_1 = 3.832 - \frac{\rho \sin(y_1)}{0.105 y_1^2 - 0.783 \rho \sin(y_1)}, \quad (5.3)$$

where y_1 is now $y_1^2 = \rho^2 + z_0^2$. The value $z_0 = 3.832$ corresponds to the first Fraunhofer diffraction minimum. Comparison of the positions of the first minima predicted by various formulas (diffraction formulas 5.1 and 5.3) against Mie predictions for $20 \leq x \leq 100$ are shown in Figure 5.2. Numerical comparison reveals that the predictions of (5.1) and (5.3) are equally close to exact results.

Following the above method, a relationship between the diameter and the position of the first minimum in the scattering pattern has been obtained by Sharma et al. (1997a) for an infinitely long cylinder as well. This relation reads as:

$$2a \sin \theta_1 = \left[1 + \frac{\pi \rho}{2 y_1} Y_1(y_1)\right] \lambda, \quad (5.4)$$

where y_1 is now $y_1 = \sqrt{\rho^2 + \pi^2}$. Predictions of (5.4) have been found to be in good agreement with exact results, particularly for large ρ. For completeness, the minima predicted by the Fraunhofer diffraction were also examined.

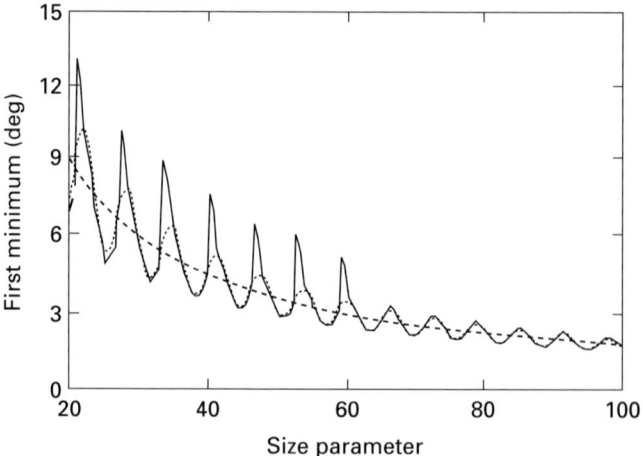

Figure 5.2. Locations of the first minimum *versus* size parameter for $m = 1.50$ for an infinitely long homogeneous cylinder: full curve, exact results; broken curve, predictions of (5.4); dotted curve, predictions of Fraunhofer diffraction formula.
From Sharma et al. (1997a).

The formation and migration of extrema with size, ellipticity and refractive index has been examined by Shepelevich et al. (1999) within the framework of the Wentzel–Kramers–Brillouin approximation (WKBA) for a nonabsorbing soft spheroidal particle. The starting point is a formula which is an analog of (3.90) for a spheroid:

$$S(\theta)_{WKB} = \frac{ix^2(m^2-1)}{m-\cos\theta} \times \int_0^1 J_0\left(x\sin\theta\sqrt{1-t^2}\right) \sin[ex(m-\cos\theta)t] \exp(i\rho t/2) t \, dt, \tag{5.5}$$

where $\rho = 2ex(m-1)$, $e = b/a$ and $2a$ is the dimension along the major axis. It can be seen that for $\theta = 0$ and $\epsilon = 1$, the above equation is the same as (3.90). Then, under the condition that $x \gg 1$ and $\theta > e(m-1)L$, where L is an empirical quantity that changes from $L \sim 0.34$ for $e \leq 1$ to $L \sim 0.17$ for $e \gg 1$, it can easily be shown that the position of the pth minima in the scattered intensity pattern is given by the relation:

$$U_1 + U_2 = \pi + 2\pi p, \tag{5.6}$$

and the position of the pth maxima in the scattered intensity pattern is determined by the relationship:

$$U_1 + U_2 = 2\pi + 2\pi p. \tag{5.7}$$

where

$$U_1 = \sqrt{x^2 \sin^2\theta + B_1^2}, \quad U_2 = \sqrt{x^2 \sin^2\theta + B_2^2}, \tag{5.8}$$

and

$$B_1 = \rho/2 + x\,e(m - \cos\theta), \quad B_2 = x\,e(\cos\theta - 1). \tag{5.9}$$

Numerical study of the migration of extrema reveals that the location of different minima migrates monotonically with the size of the spheroid. But, unlike the location of the minima, the location of the maxima moves to smaller angles in an oscillating manner.

On the basis of the above analysis, the separation between the first and the third minima can be easily related to the diameter of a spherical particle as follows:

$$d \cong \frac{54.4}{\Delta\theta_3(20)}. \tag{5.10}$$

It has been assumed in arriving at the above equation that $\sin(\theta/2) = \theta/2$.

It was noted in Chapter 3 that the eikonal picture (EP) also reproduces the positions of maxima and minima in the scattering pattern very accurately. This fact, therefore, could be exploited to obtain relationships similar to (5.6) and (5.7). The only change is that B_1 and B_2 now need to be re-defined as:

$$B_1 = e\rho_{EA}/2 + xeU; \quad B_2 = e\rho_{EA}/2 - xeU,$$

where U is represented by equation (3.89). This change, however, does not result in any significant difference from the results derived using (5.9).

Empirical formulas describing the distance between the first and the jth minima $\Delta\theta_j(\phi_d)$ – that occur after the boundary angle ϕ_d – have been obtained by Chernyshev et al. (1995) and Maltsev et al. (1996) for certain domains of particle size and refractive indexes for Mie scatterers. For $\Delta\theta_3(20)$, in the domain $m = 1.05$–1.15, $d = 1$–$12\,\mu m$, $n_0 = 1.333$ (refractive index of the medium surrounding the scatterers) and $\lambda = 632.8\,nm$ (wavelength in vacuum), d is given in (Shepelevich et al., 1999):

$$d = C_1 + C_2[\Delta\theta_3(20)]^{-2} + C_3[\Delta\theta_3(20)]^{-3} + C_4[\Delta\theta_3(20)]^{-4}, \tag{5.11}$$

where $C_1 = 0.127$, $C_2 = 52.4$, $C_3 = 190$ and $C_4 = -660$. The argument of θ is the angle in degrees.

Comparison of the predictions of equations (5.10) and (5.11) shows that (5.11) reproduces the main term of (5.10) with an error less than 4%. The error in approximating $\sin(\theta/2)$ by $\theta/2$ in the angle range 20–40 degrees is less than 2%.

Another formula obtained for $\Delta\theta_2(20)$, in the domain $m = 1.125$–1.20, $d = 1$–$10.6\,\mu m$, $n_0 = 1.333$ and $\lambda = 632.8\,nm$ is given by:

$$d = C_1 + C_2[\Delta\theta_2(20)]^{-2} + C_3[\Delta\theta_2(20)]^{-3} + C_4[\Delta\theta_2(20)]^{-4}, \tag{5.12}$$

where $C_1 = -0.06$, $C_2 = 28.8$, $C_3 = 600$ and $C_4 = -1,700$.

An equation that relates m to the positions of maxima and minima has also been given by Chernyshev et al. (1995). For this, one may define a quantity termed "visibility" as:

$$V(\phi_V) = \frac{I_{max} - I_{min}}{I_{max} + I_{min}}, \tag{5.13}$$

where I_{min} is the intensity at the angle of next minimum after the boundary angle ϕ_V. Similarly, I_{max} is the first maximum after the boundary angle. The equation that relates m to visibility is:

$$m = \frac{1}{n_0}[D_1 + D_2[V(30)]^{1/2} + D_3 V(30) + D_4[V(30)]^2], \qquad (5.14)$$

where $D_1 = 3.9$, $D_2 = -10$, $D_3 = 12$ and $D_4 = -5$. This equation yields a standard error of 0.03 for particle refractive index calculation over the range mentioned for the validity of (5.12).

A comparison of errors in fibre size determination from the positions of minima in a scattering pattern was carried out by Sharma and Somerford (1988) for intermediate size (diameter) nonabsorbing fibres. The approximations considered were the RGDA and two of its modified versions – namely, RGDA1 (Shimizu, 1983) and RGDA2 (Gordon, 1985). For numerical comparison, the errors in size determination from FCI were also included. The minima in the RGDA, RGDA1 and RGDA2 are determined by the zeros of $J_1(U)/U$. The first three zeros occur at $U = 3.832$, 7.016 and 10.173. The relationships for the three approximations are:

$$U = 2x \sin(\theta/2), \quad \text{for RGDA} \qquad (5.15)$$

$$U = 2mx \sin(\theta/2) \quad \text{for RGDA1,} \qquad (5.16)$$

and

$$U = x(1 + m^2 + 2m \cos \theta)^{1/2} \quad \text{for RGDA2.} \qquad (5.17)$$

Typical percent errors are shown in Table 5.3. It is clear that neither of the modified forms of the RGDA give better estimates of fibre size than those obtained by the use of FCI. Nevertheless, it is clear that the RGDA2 constitutes a significant improvement on the RGDA – at least as far as the positions of the extrema are concerned – and can be used with reasonable accuracy in a straightforward and simple way. But, although the simple relationship between the positions of minima and the size in the RGDA2 makes the method much faster, the measurements in the scattering pattern have to be

Table 5.3. Percent error in size determination from the position of the first minimum using RGDA, RGDA1, and RGDA2 for $n = 1.05$,* and percent errors in the FCI.**

x	RGDA	RGDA1	RGDA2	FCI
2.0	−4.6	0.4	−2.05	0.60
3.0	−3.4	1.53	0.83	0.38
5.0	−2.96	1.94	0.26	0.27
7.0	−2.74	2.14	0.14	0.19
9.0	−2.81	2.08	0.35	0.17
11.0	−2.85	2.05	0.65	0.15
13.0	−3.33	1.58	0.60	0.14
15.0	−3.34	1.58	1.06	0.12
20.0	−4.70	0.29	1.28	0.10

*From Sharma and Somerford (1988).
**From Sharma and Somerford (1983a).

made over a much wider angular range. The first minimum for $m = 1.05$ and $x = 2.0$, for example, occurs at $132.7°$.

The size parameter domain of the validity of these approximations depends on which minimum is chosen to deduce the size of the scatterer. For size determination of an infinitely long cylinder at perpendicular incidence from first, second and third minima, respectively, the domains of validity are $2 \leq x \leq 20$, $5 \leq x \leq 25$ and $7 \leq x \leq 30$. The errors in size determination are than less than 3%. Formulas (5.15) to (5.17) also relate the position of extrema to the size of a sphere too. The first four minima for a sphere occur at $U = 4.493, 7.73, 10.90$ and 14.08.

For a spheroid Barber and Wang (1978) have examined the change in the region of applicability of the RGDA as a sphere deviates to a spheroid. They found that the error increases with $x = ka$ and decreases with axial ratio $a : b$ for small particles, but increases with axial ratio for large particles. It was also found that the worst case occurs when the incident wave is along the major dimension (a) of the spheroid. The relation between the position of minima and the size is (see, e.g., Barber and Wang, 1978):

$$U = 2x \sin \frac{\theta}{2} \left(\cos \beta + \frac{b^2}{a^2} \sin^2 \beta \right)^{1/2}. \tag{5.18}$$

The angle β is the angle between the axis of the ellipsoid and a line bisecting the incident and observation angles. In the event that the incident radiation is perpendicular to the semi-major axis a, the angle $\beta = \theta/2$. Thus, the above relation becomes:

$$U = 2x \sin \frac{\theta}{2} \left(\cos(\theta/2) + \frac{b^2}{a^2} \sin^2(\theta/2) \right)^{1/2}. \tag{5.19}$$

An estimation of errors in determining size from (5.19) has been done by Yadav and Sharma (unpublished). The results obtained from an analogy to RGDA1 and WKBA were also included in the comparison. It was concluded that for small deviations from sphericity, the formula based on the WKBA turns out to be most accurate. However, in contrast to what was observed by Barber and Wang (1978), it has been noted that – as the departure from sphericity increases – the error in determining size based on RGDA-type approximations decreases. This is the result of the fact that – while the difference between is exact and the RGDA-type scattering pattern increases as the deviation from sphericity increases – the positions of the extrema continue to be reproduced correctly.

5.1.2 Suspension of particles

For a monodisperse system of spherical particles of concentration c the experimentally observed quantity $I(\theta)/I_0 c$ is related to $i(\theta)$ for a single particle through relations (Heller et al., 1959):

$$\frac{I(\theta)}{I_0 c} = \frac{\lambda^2 i(\theta)}{4\pi^2 w_t}, \tag{5.20}$$

and

$$\frac{I(\theta)}{I_0 c} = \frac{3g_{12}\lambda^2 i(\theta)}{2g_2 \pi^3 d^3}, \tag{5.21}$$

where $I(\theta)/I_0 c$ is the specific intensity of light scattered at an angle θ and g_2 and g_{12} are the densities of the particles and the total system. Particle weight and diameter are denoted by w_t and d, respectively. Since the EA gives a simple analytic expression for $i(0)$, one may obtain the particle size from (5.21) by checking the value of x that satisfies (5.21). The value of w_t then follows from (5.20). The percent error introduced when determining weight and size by using $i(0)_{EA}$ in place of exact $i(0)$ has been estimated and compared with the percent error introduced by using the RGDA and the ADA (Sharma and Debi, 1980). The maximum percent error in using the EA for determining weight and size in the domains $1 \leq x \leq 25$ and $m \leq 1.05$ is of the order 9% and 3%, respectively.

An enormous amount of work has been done on methods for obtaining particle size distribution from the extinction spectrum or the optical turbidity. The extinction coefficient $K_{ext}(m, k)$ for light scattered by a dilute suspension of polydisperse particles of similar optical properties is given by the relation (see, e.g., Bayvel and Jones, 1981):

$$K_{ext}(m, k) = 2\pi N \int Q_{ext}(m, ka) a^2 f(a)\, da, \tag{5.22}$$

where N is the number of particles per unit volume in suspension and $Q_{ext}(m, ka)$ is the extinction efficiency factor of a single particle when the light of wave number k (in the medium of suspension) is scattered by a particle of radius a and refractive index m (relative to medium). The probability of obtaining a particle in the size range a and $a + da$ in the unit volume is $f(a)\, da$.

Exact solutions of the Fredholm integral equation (5.22) are difficult and not always possible to obtain. The simplest and perhaps most widely used method is to find an empirical distribution that satisfies (5.22). However, this method does not guarantee closeness of $f(a)$ to actual $f(a)$. An analytic inversion of (5.22) is possible if exact Q_{ext} appearing as a kernel function is replaced by a suitable approximate form. Shifrin and Perelman (1967) used this idea to invert the light scattering data of soft particles using the ADA. Analytic inversion methods based on this approximation have also been derived by Fymat (1978) and Box and McKeller (1978) for spheres; McKellar (1982) for infinitely long cylinders; Smith (1982) for particles with a variable complex refractive index; and Punina and Perelman (1969) and Klett (1984) for absorbing spheres. A derivation based on the complex analytic extension of the ADA for analytic inversion of spectral extinction data has been presented by Franssens et al. (2000). The derivation is applicable to homogeneous nonabsorbing as well as absorbing spherical particles. The procedure generalizes and unifies a number of results obtained previously. We will not discuss this class of methods in detail. An excellent review of the topic has been given by Shifrin and Tonna (1993). Bayvel and Jones (1981) have also treated the subject at length.

Use of the above-mentioned analytic inversion methods requires *a priori* knowledge of the total surface area of spheres and/or the total number of spheres. However, these conditions are sometimes impossible to fulfill in practical situations. An analytic inversion method for spherical particles based on the ADA has been developed by Wang and Hallett (1996). This approach eliminates the need for *a priori* knowledge of the total number of particles and their total area. It is assumed that $f(a)$ is absolutely integrable, has finite extrema and that particles are nonabsorbing. Substituting the expression Q_{ADA}^{ext} for Q_{ext} in (5.22), one obtains:

$$2k^2 K_{ext}(m,k) = \pi N \int_0^\infty [2(2ka)^2 - 4(2ka)\sin(2ka) + 4 - 4\cos(2ka)]f(a)\,da. \quad (5.23)$$

Next, it follows, from the use of the Riemann–Lebesgue lemma, that:

$$\int_0^\infty a^2 f(a) \cos(2ka)\,da = \frac{1}{4\pi N}\left[\lim_{k\to\infty}\left(\frac{d(k^2 K_{ext})}{k\,dk}\right) - \frac{d(k^2 K_{ext})}{k\,dk}\right]. \quad (5.24)$$

Then, size distribution can be inverted through a Fourier cosine transform:

$$f(a) = \frac{1}{2a^2}\int_0^\infty \left[\lim_{k\to\infty}\frac{d(k^2 K_{ext})}{k\,dk} - \frac{d(k^2 K_{ext})}{k\,dk}\right]\cos(2ka)\,dk. \quad (5.25)$$

Equation (5.25) constitutes the basis for inverting the size distribution from extinction data through the fast Fourier transform. Numerical tests performed by Wang and Hallett (1996) show that this analytic inversion procedure offers a simple and fast solution to accurate inversion of size distributions from the turbidity spectrum of nonabsorbing spheres.

A distinct approach that is based on the mean value theorem and the method of Lagrange multipliers has been developed by Roy and Sharma (1997). This procedure is applicable to a collection of Mie particles described by a single modal distribution function of moderate skewness. This means that the distribution is skew to the extent that $f(a)$ could be taken to be a continuous function over the whole size range. The ADA has been employed in this scheme for expression of the extinction efficiency for a single particle. It has been demonstrated that the key parameters that characterize the particle size distribution function can be obtained by repeated application of the first mean value theorem of the integral calculus to (5.22). The key parameters so obtained are (*i*) minimum and maximum particle radii with assistance from the ADA, (*ii*) particle number, (*iii*) mean radius, (*iv*) mean square radius, (*v*) mean cube radius and (*vi*) mean fourth-power radius. Once these parameters are known, the mode, median, skewness, kurtosis, etc. can easily be evaluated. To show how the mean value theorem works for determination of these parameters, we have described its use in Appendix D. The standard Pearson method may then be used to obtain the distribution profile (Appendix E). The value of a_0 (the minimum size) is taken to be 0 in these calculations. This results in obtaining a slightly deformed distribution profile near the origin. But, this is a common feature of most existing inversion methods as well. Generally, it is the

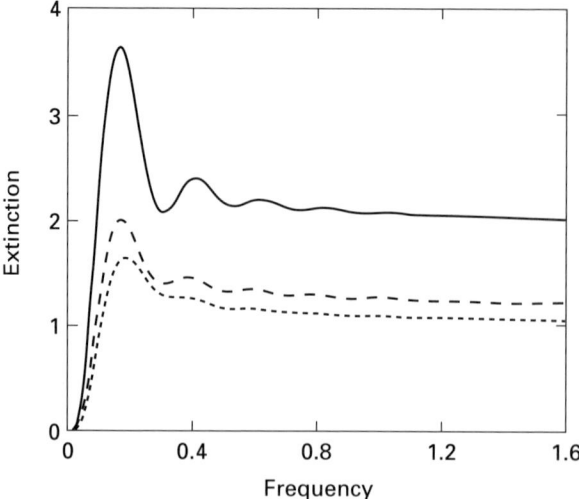

Figure 5.3. Typical variations of extinction efficiency factor with size parameter and relative refractive index.

value of a_m (the maximum radius) that is considered to be the important information one tries to obtain through study of extinction spectra. To that extent, this approach has been found to yield satisfactory values in a case study examined by Roy and Sharma (1997). Another direct method involves choosing an empirical distribution function in keeping with the basic features and subjecting it to the required number of conditions as would be needed to evaluate the parameters occurring in the empirical formula. These methods of obtaining the distribution profile, however, are not based on the formal solution of (5.22). A formal solution can be constructed using the Lagrange multiplier method. The expression for $f(a)$ obtained by the Lagrange multiplier method should actually be regarded as an ansatz solution for $f(a)$ given by (5.22) and requires knowledge about all orders of moments. But, the fact that only a finite number of moments can be used in a real situation restricts the accuracy of the method. However, knowledge of the first four moments has been shown to give reasonably good results.

In yet another solution to the problem, Roy and Sharma (2005) have examined the extinction spectrum generated by a cloud of soft Mie particles of smooth size distributions. It is shown that the extinction spectrum, in general, has some easily identifiable characteristic regions where the extinction–frequency relationship can be approximated by simple empirical formulas involving the first four moments of the particle size distribution function. Figure 5.3 shows typical graphs of $K_{ext}(m,\nu)$ versus $\nu(=k/2\pi)$ over a wide range of ν with some prescribed values of N, m and three definitive forms of $f(a)$ – namely, (*i*) Gaussian, (*ii*) beta and (*iii*) gamma. The variational pattern of $K_{ext}(m,\nu)$ with respect to ν has some invariant features which may be identified as follows:

1. For very small values of ν, $K_{ext}(m,\nu)$ is almost 0 irrespective of the nature of $f(a)$ and its range $[a_0, a_m]$. This is expected, as we know that $Q_{ext}(m, \nu a) \to 0$ as $\nu a \to 0$. This region, however, is of very little interest from a practical viewpoint.

2. As ν increases, $K_{ext}(m,\nu)$ starts rising. At first, it rises slowly and then almost at a uniform rate. In the region where $K_{ext}(m,\nu)$ has almost linear growth, the rate of growth depends very much on the nature and the range of $f(a)$. This linear growth may be conveniently represented as:

$$K_{ext}(m,\nu) = 2\pi N[A\nu - B], \qquad (5.26)$$

where A and B are constants. A comparison of the right-hand sides of (5.26) and (5.22) suggests that $K_{ext}(m,\nu)$ in this region may, in fact, be written as:

$$K_{ext}(m,\nu) = 2\pi N\left[l_1^+ \nu \bar{a}^3 - l_2^+ \bar{a}^2\right], \qquad (5.27)$$

where the two designated coefficients l_1^+ and l_2^+ depend only on m.

3. With a further increase of ν, this region of linear rise of $K_{ext}(m,\nu)$ gradually begins to show a tendency to bend forward in a continuous and smooth manner. As a result, $K_{ext}(m,\nu)$ eventually reaches its first maximum value (K_{max}^1) for some $\nu (= \nu_{max}^1$, say) and then – with an even further increase in ν – $K_{ext}(m,\nu)$ starts a smooth downward journey. We get an approximately inverted parabolic trajectory near $\nu = \nu_{max}^1$. This suggests that near $\nu = \nu_{max}^1$, $K_{ext}(m,\nu)$ assumes the well-known representative form:

$$K_{ext}(m,\nu) = K_{max}^1 - \alpha \frac{(\nu - \nu_{max}^1)^2}{2}. \qquad (5.28)$$

The slope $\partial K_{ext}/\partial \nu$ is 0 at $\nu = \nu_{max}^1$, positive for $\nu < \nu_{max}^1$ and negative for $\nu > \nu_{max}^1$. The ν region for which K_{ext} shows this parabolic behavior varies sharply with the range and form of $f(a)$. Comparison of (5.28) and (5.22) immediately suggests that we can write (5.28) as:

$$K_{ext}(m,\nu) = 2\pi N\left[-p_1^+ \nu^2 \bar{a}^4 + p_2^+ \nu \bar{a}^3 - p_3^+ \bar{a}^2\right], \qquad (5.29)$$

where p_1^+, p_2^+ and p_3^+ are functions of m only. The following relations are easily identified:

$$K_{max}^1 = 2\pi N\left[\frac{(p_2^+ \bar{a}^3)^2}{4 p_1^+ \bar{a}^4} - p_3^+ \bar{a}^2\right], \qquad (5.30)$$

$$\alpha = 4\pi N p_1^+ \bar{a}^4, \qquad (5.31)$$

and

$$\nu_{max}^1 = \frac{p_2^+}{p_1^+} \frac{\bar{a}^3}{2\bar{a}^4}. \qquad (5.32)$$

when (5.29) is compared with (5.28).

4. As ν increases further, $K_{ext}(m,\nu)$ shows a nearly linear decline. Using similar arguments to those for linear increase, we can write in this region:

$$K_{ext}(m,\nu) = 2\pi N\left[-l_1^-\overline{\nu a^3} + l_2^-\overline{a^2}\right], \tag{5.33}$$

where l_1^-, l_2^- are functions of m only.

5. After a nearly linear decline, the rate of decline falls slowly and continuously so that the first minimum of K_{ext} occurs at the point $\nu = \nu_{min}^1$ where the slope $\partial K_{ext}(m,\nu)/\partial \nu$ is horizontal. The representative form of $K_{ext}(m,\nu)$ near $\nu = \nu_{min}^1$ may be taken as:

$$K_{ext}(m,\nu) = K_{min}^1 + \alpha' \frac{(\nu - \nu_{min}^1)^2}{2}. \tag{5.34}$$

This form ensures that K_{min}^1 is indeed the minimum value of $K_{ext}(m,\nu)$ at $\nu = \nu_{min}^1$ and the slope $\partial K_{ext}(m,\nu)/\partial \nu$ is negative for $\nu < \nu_{min}^1$ and positive for $\nu > \nu_{min}^1$. Following equations (5.29) to (5.32), the expression for $K_{ext}(m,\nu)$ near $\nu = \nu_{min}^1$ is:

$$K_{ext}(m,\nu) = 2\pi N\left[p_1^-\overline{\nu^2 a^4} - p_2^-\overline{\nu a^3} + p_3^-\overline{a^2}\right] \tag{5.35}$$

and, consequently, we have the following relations:

$$K_{min}^1 = 2\pi N\left[-\frac{(p_2^-\overline{a^3})^2}{4p_1^-\overline{a^4}} + p_3^-\overline{a^2}\right], \tag{5.36}$$

$$\alpha' = 4\pi N p_1^-\overline{a^4}, \tag{5.37}$$

and

$$\nu_{min}^1 = \frac{p_2^-}{p_1^-}\frac{\overline{a^3}}{2\overline{a^4}}. \tag{5.38}$$

The coefficients p_1^-, p_2^-, p_3^- depend only on m.

6. After the first minimum, $K_{ext}(m,\nu)$ passes through some more maxima and minima in a regular fashion with the amplitude of oscillations decreasing rapidly. For very large ν and a very large value of a_m, $K_{ext}(m,\nu)$ falls off continuously with a small, varying negative slope and asymptotically approaches a constant value. This behavior of $K_{ext}(m,\nu)$ suggests an appropriate representative form:

$$K_{ext}(m,\nu) = 2\pi N\left[\frac{l_1^\infty}{\nu}\bar{a} + l_2^\infty\overline{a^2}\right], \tag{5.39}$$

where l_1^∞, l_2^∞ are coefficients which depend on m.

These are the six noticeable regions which are very easily observed in an extinction spectrum covering the entire ν domain. In practice, however, all six regions cannot be realized at a time in an experimental situation. But, with some care one can arrange to have sufficient $K_{ext}(m,\nu)$ data to generate at least one of the above-mentioned six regions. One can then employ the appropriate functional form for $K_{ext}(m,\nu)$ corresponding to the region for further analysis.

Table 5.4. Verification of formulas (5.30), (5.32), (5.36) and (5.38) for the locations and magnitudes of the first maximum and the first minimum of K_{ext}.

Distribution type	K^1_{max} (5.30)	K^1_{max} Fig. 5.3	ν^1_{max} (5.32)	ν^1_{max} Fig. 5.3	K^1_{min} (5.36)	K^1_{min} Fig. 5.3	ν^1_{min} (5.38)	ν^1_{min} Fig. 5.3
Beta	3.622	3.647	0.170	0.168	1.83	2.08	0.310	0.307
Gaussian	2.018	2.007	0.175	0.170	1.403	1.401	0.320	0.304
Gamma	1.675	1.648	0.187	0.190	–	–	–	–

From Roy and Sharma (2005).

The five functional forms of $K_{ext}(m, \nu)$ – namely, equations (5.27), (5.29), (5.33), (5.35) and (5.39) expressed above – contain $\bar{a}, \overline{a^2}, \overline{a^3}, \overline{a^4}$ as variable lumped parameters which are characteristics of the distribution function $f(a)$. This contains 12 designated coefficients $l_1^+, l_2^+, p_1^+, p_2^+, p_3^+, l_1^-, l_2^-, p_1^-, p_2^-, p_3^-, l_1^\infty$ and l_2^∞. This set of coefficients is not dependent on the nature of $f(a)$ or its range. However, it does depend on m. Hence, one may use some test distributions (normalized prefixed distribution functions) – namely, Gaussian, gamma, beta, Weibüll, etc. – so that the lumped parameters $\overline{a^2}, \overline{a^3}, \overline{a^4}$ are a priori known. One can then generate extinction spectra numerically for any desired value of m, and the set of corresponding coefficients can be evaluated from there. Insertion of these coefficients in the functional forms of the $K_{ext}(m, \nu)$ results in having formulas (valid in the respective ν regions) that are useful for extracting information about particle size distribution (i.e., estimation of $\overline{a^2}, \overline{a^3}, \overline{a^4}$ and hence an approximated reproduction of $f(a)$ if the range is given) from knowledge of the extinction spectrum in a given situation with a specification of N, the particle number and m the relative refractive index. The correctness of the formulas has been firmly established in the detailed analysis by Roy and Sharma (2005). As an example, in Table 5.4 we give the locations and magnitudes of the first maximum and the first minimum obtained from the approximate formulas and from the exact calculations.

Lui et al. (1996) have obtained a closed form approximation for (5.22) based on replacing the Q_{ext} by Q_{ADA}^{ext} and a Deirmendjian modified gamma function form for the particle size distribution (Deirmendjian, 1969):

$$f(a) = C a^\alpha e^{-ua^\gamma}, \qquad (5.40)$$

where C is the normalization constant defined as:

$$N = \int f(a)\, da = \frac{C u^{-(\alpha+1)/\gamma}}{\gamma} \Gamma\left(\frac{\alpha+1}{\gamma}\right), \qquad (5.41)$$

with N as the number density. It can be shown that Q_{ADA}^{ext} for a sphere of almost real refractive index can be expanded in a series as (Nicholls, 1984):

$$Q_{ADA}^{ext} = \sum_{i=0}^{\infty} \frac{q_i a^{2i}}{\lambda^{2i}}, \qquad (5.42)$$

where

$$q_i = \frac{4(-1)^{i+1}(2i+1)[4\pi(m-1)]^{2i}}{(2i+2)!}. \tag{5.43}$$

Thus the ith term in the series is of the order ρ^{2i}. Using this expansion, it can be verified that K_{ADA}^{ext} can be written as (Lui et al., 1996):

$$K_{ADA}^{ext}(m,k) = \frac{\pi C}{\gamma}\sum_{i=1}^{\infty} A_i, \tag{5.44}$$

where

$$A_i = \frac{q_i \Gamma(y)}{b^y \lambda^{2i}}, \quad \text{and} \quad y = \frac{\alpha + 2i + 3}{\gamma}. \tag{5.45}$$

Equations (5.44) and (5.45) have been shown to be good analytic approximations for atmospheric aerosols containing water droplets.

For (5.44) to be a good approximation, the series should converge. In other words, the inequality:

$$\mathcal{L} = \lim_{i \to \infty} \left|\frac{A_{i+1}}{A_i}\right| < 1,$$

must be satisfied. This yields the following restrictions on γ:

(i) For $\gamma < 1$, the series diverges irrespective of the values of u, α, γ and λ.
(ii) For $\gamma = 1$:

$$\mathcal{L} \equiv g = \frac{4\pi(m-1)}{u\lambda}.$$

If $g > 1$ the series diverges and if $g < 1$ the series converges. For $g = 1$, the series fails. The critical frequency is $u/(4\pi(m-1))$, below which the series converges.
(iii) For $\gamma > 1$, $\mathcal{L} = 0$ for all $u, \alpha, \gamma, \lambda$ and the series always converges.

In cases (ii) and (iii), tests are necessary to discover the frequency range in which the series converges to a limit comparable with that obtained by Mie calculation (discrepancy <20%). Numerical comparisons have been made by Lui et al. (1996) between approximate K_{ext} and numerical evaluation of the exact Mie theory. Agreement is excellent over frequency domains corresponding to the ultraviolet, optical and infrared spectral regions for modified γ distributions.

It is well-known that the real and imaginary parts of the refractive index, n_d and n'_d, respectively, of a dilute dispersion can be written as (see, e.g., van de Hulst 1957):

$$n_d = m_1 + 2\pi m_1 k^{-3} N \operatorname{Im}[S(0)], \tag{5.46}$$

and

$$n'_d = 2\pi m_1 k^{-3} N \operatorname{Re}[S(0)], \tag{5.47}$$

where m_1 is the refractive index of the continuous phase, N is the number of particles per unit volume and k is the wave number in the medium. For a dispersion of spherical particles the volume fraction is:

$$v_f = \frac{4\pi a^3}{3} N, \tag{5.48}$$

where a is the radius of particles. If we denote by m_2 the refractive index of each particle immersed in the medium, then equations (5.46) and (5.47) in terms of volume fraction, for real m_1 and m_2, can be expressed as:

$$n_d = m_1 + \frac{3m_1 v_f}{8} \frac{P_{sca}(m_2, x)}{x}, \tag{5.49}$$

and

$$n'_d = \frac{3m_1 v_f}{8} \frac{Q_{sca}(m_2, x)}{x}, \tag{5.50}$$

where P_{sca} is the refraction analog of extinction efficiency Q_{sca} and is given by the relation:

$$P_{sca} = \frac{4}{x^2} \text{Im}[S(0)]. \tag{5.51}$$

Further, the experimentally measured quantity – that is, the rate of change of the real and imaginary parts of refractive index of dispersion with concentration – can be readily expressed as:

$$\frac{dn_d}{dv_f} = \frac{3m_1}{8\bar{\rho}_2} \frac{P_{sca}}{x}, \tag{5.52}$$

and

$$\frac{dn'_d}{dv_f} = \frac{3m_1}{8\bar{\rho}_2} \frac{Q_{sca}}{x}, \tag{5.53}$$

where v_f is the weight concentration of particles of refractive index m_2 and $\bar{\rho}_2$ is the density.

Zimm and Dandlikar (1954) and Champion et al. (1979) have studied the dependence of the refractive index of a colloidal dispersion on the size of the dispersed particles. It has been shown theoretically that for spheres of size of the order of wavelength, the refractive index of the dispersion is strongly size-dependent and, hence, is a useful sizing parameter. A general result of the ADA as well as the Mie theory is that the specific refractive index increment dn_d/dv_f of a dispersion of spheres tends to 0 in an oscillatory manner as the sphere diameter increases. Also, for particles that are small compared with the wavelength of incident light, the refractive index of the dispersion is independent of size but is sensitive to the volume fraction of the dispersed material. Hence, n_d is a useful indicator for the measurement of total dispersed phase concentration.

Meeten (1980b) has extended this study to examine the effect of the shape of the particle on the refractive index of the dispersion. The real refractive index of an ensemble of nonspherical particles can be written as:

$$n_d = m_1 + 2\pi m_1 N k^{-3} \langle \text{Im } S(0) \rangle, \tag{5.54}$$

where the angular braces indicate that an orientation average has been taken over all possible orientations of the scattering particle. But for this difference, the above equation is the same as (5.46). For a general ellipsoid of semi-axes a, b, c this may be written as:

$$n_d = m_1 + \frac{3}{2} m_1 v_f \frac{\langle \text{Im } S(0) \rangle}{x_a x_b x_c}, \quad (5.55)$$

where $x_a = ka$, etc. are analogous to the usually defined size parameter. The specific refractive index increment of the dispersion then yields:

$$\frac{1}{m_1} \frac{dn_d}{dv_2} = \frac{3}{2} \frac{\langle \text{Im } S(0) \rangle}{x_a x_b x_c}. \quad (5.56)$$

The right-hand side of the above equation has been calculated by Meeten (1980b) using the ADA. The dependence of $(1/m_1)(dn_d/dv_f)$ on the volume-averaged particle size parameter defined as $(x_a x_b^2)^{1/3}$ has been studied for various $a:b$ ratios. It is found that for particles of different shape, but equal volume, the oscillatory dependence of dn_d/dv_2 shown by the spheres disappears as the deviation from spherical shape increases. It is concluded that refraction, like turbidity, can be used to infer particle size and shape. Meeten (1979) has also studied the effect of shape of the particle on induced birefringence and optical anisotropy in a colloidal dispersion in the context of kaolinite clay particles which are hexagonal plate-like particles. Results show that the ADA is capable of explaining the experimental particle size dependence for kaolinite particles.

For a uniaxial birefringent particle, it is well-known that the refractive index may be expressed as:

$$\begin{pmatrix} m_e & 0 & 0 \\ 0 & m_o & 0 \\ 0 & 0 & m_o \end{pmatrix},$$

where m_e is the refractive index as seen by the extraordinary ray and m_o is the refractive index as seen by the ordinary ray. The complex value of linear birefringence is then:

$$\Delta m = 2\pi m_1 N k_{-3} \left[S(0)_e - S(0)_o \right]. \quad (5.57)$$

The subscripts o and e refer to the ordinary and extraordinary rays, respectively. Thus, for calculation of complex birefringence what is required is knowledge of $(S(0)_e - S(0)_o)$. This can be easily calculated for a variety of scatterers using soft-particle approximations. Meeten (1979) has obtained an appropriate expression in the ADA for linear birefringence and studied it numerically for a colloidal dispersion of kaolinite particles. These studies predict strong particle size dependence for wavelength-sized colloidal particles. This observation contradicts the results of Peterlin and Stuart (1939) which predict independence of birefringence on particle size. If the scatterers are weakly birefringent spheres, then with the help of (3.39) equation (5.57) can be rewritten as:

$$\Delta m = v_f (m_e - m_o)(\mathcal{B}'(\tilde{\rho}) + i\mathcal{B}''(\tilde{\rho})), \quad (5.58)$$

where

$$\mathcal{B}'(\tilde{\rho}) = \frac{3}{\tilde{\rho}}\sin(\tilde{\rho}) + \frac{6}{\tilde{\rho}^2}\cos(\tilde{\rho}) - \frac{6}{\tilde{\rho}^3}\sin(\tilde{\rho}), \tag{5.59}$$

is the linear birefringence function and:

$$\mathcal{B}''(\tilde{\rho}) = \frac{3}{\tilde{\rho}}\cos(\tilde{\rho}) + \frac{6}{\tilde{\rho}^2}\sin(\tilde{\rho}) - \frac{6}{\tilde{\rho}^3}(1 - \cos(\tilde{\rho})), \tag{5.60}$$

is the linear dichroism function $\tilde{\rho}$ defined as $\tilde{\rho} = x(m_e + m_o - 2)$. The linear birefringence function describes the phase difference between linearly and orthogonally polarized light, and the linear dichroism function describes the relative change of amplitude. Numerical evaluation of variation of $\mathcal{B}'$ and $\mathcal{B}''$ with $\tilde{\rho}$ and particle radius a shows that the maxima and minima of the linear dichroism function correspond closely with the zeros of the birefringence function (Meeten, 1980a). This is in accordance with Zocher's rule (1925) which states that the linear dichroism of a colloidal dispersion reaches a maximum when its linear birefringence is 0.

For small and weakly refracting particles $\mathcal{B}' \to 1$ and $\mathcal{B}'' \to 0$. This gives:

$$\Delta m = v_f(m_e - m_o),$$

which is the well-known result of Peterlin and Stuart (1939). This means that for small particles the results of the ADA agree with Peterlin and Stuart's expressions.

Borovoi and Krutikov (1976) have calculated, within the framework of the EA, the statistical characteristics of the wave field propagating in the polydispersion of weakly refractive homogeneous spheres and have shown that the measured statistical characteristics can be used to determine the average characteristics of individual particles.

5.1.3 Aggregates

Small particles dispersed in a fluid are commonly known to aggregate. Because of the random nature of agglomerate it has become popular to treat them as fractals. A description of light scattering by such particles has been given by Berry and Percival (1986). However, it has been noted that their theory does not give the proper asymptotic behavior of efficiencies in the asymptotic limit $M \to \infty$, where M is the cluster mass.

Khlebtsov (1993) has proposed a theory of scattering by fractal clusters which is based on the ADA. In this model it is assumed that the fractal cluster can be treated as a porous sphere of spherically symmetric mass density distribution of the form:

$$d(r) \sim \left(\frac{r}{a}\right)^{D-3} h(r/R), \tag{5.61}$$

with $h(r/R)$ is the cut-off function and D is the fractal dimension. Here, R is the characteristic cluster size. If the agglomerate is made up of M spherules of radius a

and complex refractive index m, the refractive index of the aggregate can be written as:

$$\tilde{m}(r) \sim 1 + \frac{m^2 - 1}{m^2 + 2} d(r), \tag{5.62}$$

by introducing the assumption that the difference between the effective refractive index $\tilde{m}$ and the surrounding medium (taken to be 1) is proportional to the density $d(r)$ and the polarizability of spherules. The assumption made in (5.62) is justified if $|\tilde{m}(r) - 1| \leq 1$. It is found that at low values of phase shift results obtained for scattering and absorption efficiencies reduce to the formulas of single-scattering theory of a large mass and cluster. For clusters with a sharp cut-off function of mass density distribution this formulation is found to lead to the correct asymptotic limits.

5.2 INTERSTELLAR AND INTERPLANETARY DUST

Analysis of interstellar extinction is a powerful tool for characterizing interstellar dust particles. The particles in these regions are almost always irregular. Measurements of microwave analog experiments (Zerull *et al.*, 1977) as well as light scattering experiments (Weiss-Wrana, 1983; Weiss, 1981) from irregular particles reveal that it is not possible to apply the smooth surface Mie theory to such particles. To evaluate the scattering from such particles the effect of surface roughness needs to be incorporated. This has been done by Chiappetta (1980) by introducing a Fermi distribution in the phase shift function. Such a procedure is physically based on the fact that the refractive index decreases continuously in a region close to the surface and is not sharply cut off. This is achieved in this model by writing the refractive index as:

$$m(r) - 1 \propto \frac{m - 1}{[1 + e^{k(b-a)}]}. \tag{5.63}$$

The predictions of this model have been found to be in good agreement with the microwave scattering measurements of Zerull *et al.* (1977) for irregular compact and fluffy particles. In theoretical calculations, the EA was used to parametrize the forward scattering function. For the backward direction the reflective model and shadow function (Wolf, 1975; Giese *et al.*, 1978) were used.

The above model – proposed by Chiappetta – was refined by Perrin and Lamy (1983). The refined model takes into account local fluctuations at the surface of the particle, the so-called "asperities". Thus, they modified the refractive index profile (5.63) to:

$$m(r) - 1 = (m - 1) \frac{1 + \exp\left[\frac{z - \sqrt{a^2 - b^2}\Theta(a - b)}{d}\right]^{-1}}{1 + \exp[(b - a)/d]}. \tag{5.64}$$

Here Θ is the step function. The model takes into account local fluctuations through d, which is the mean amplitude of fluctuations. For large-angle scattering, these authors adapted the work of Wolf (1975, 1980, 1981) on single and double reflections by the microstructures of the surface. The scattered intensity and polarization from this model were tested against experimental measurements on four different individual particles. The material type and average diameters of these particles were (a) three dielectric materials of average sizes 110 µm, 42 µm and 28 µm and (b) a metallic material of average size 20 µm. Good fits were obtained for all dielectric particles. For metallic particles, angular scattering as well as polarization were not produced well in this model. But, this is not surprising because the EA is not expected to be good for metallic particles.

Perrin and Lamy (1986) and Bourrely *et al.* (1991) have improved the above model further by incorporating the vector description in the EA. The scattering pattern and polarization obtained using a vector eikonal model for a particle from the Murchison meteorite (dielectric particle) and a particle of magnetite (metallic particle) show a definite improvement over scalar model predictions.

The phase functions of dust particles generally show a backscattering increase not observed in the phase function of a Mie particle. One possible source of this experimental observation can be the roughness at the surface of the dust particle. Bourrely *et al.* (1986a) have specifically studied the effect of roughness on backscattering in a formulation based on the EP. Formulation of the problem is straightforward. To see this, let us re-write (3.86) for an axially symmetric scatterer as:

$$S(\theta)_{EP} = -i\pi k^3 \int b\, db\, J_0(kb\sin\theta) G(b)\, db, \qquad (5.65)$$

where

$$G(\mathbf{b}) = \int [m^2(\mathbf{b},z) - 1]\, e^{-2ik\sin^2(\theta/2)}\, e^{(ik/2)\int_{\infty}^{Z(\mathbf{b})}[m^2(\mathbf{b},z')-1]dz'}, \qquad (5.66)$$

and $Z(\mathbf{b})$ denotes the height of the boundary of the scatterer for a fixed impact parameter. To account exactly for the shape of the scatterer, one can first perform integration over the z-variable for a fixed value of $\mathbf{b} = \mathbf{b}_0$. The problem is then to find the intersection points of line $\mathbf{b} = \mathbf{b}_0$ and the curve $\mathbf{b} = S(z)$, where $S(z)$ is the expression of the boundary of the scattering object. This may be done analytically for simple geometries. For instance, for a sphere the intersection gives $Z(\mathbf{b}) = \sqrt{a^2 - b^2}$. For complex scatterers, this may be done numerically.

The scattered intensities were calculated for a sphere with surface roughness modeled by a function of fractal type. The results of investigation revealed backward enhancement which was in qualitative agreement with experimental measurements of light scattered by a collection of particles (Weiss-Wrana, 1983). This feature of backscattering enhancement is, therefore, well-suited to exhibit the differences between smooth and rough surfaces. The model also predicts strong oscillations in the scattering pattern. But, these strong oscillations are not seen in experimental results because the scattering particles are not all identical in shape and, hence, oscillations are averaged out. Also, the angular resolution of the measured scattered intensity is much greater than the period of predicted oscillations. Tests have been made in the microwave range. The presence of strong oscillations and a backward scattering enhancement is observed. The angular domain 160°–180° is most sensitive to the degree of roughness of the particle. This procedure has also been used to study light scattering by other irregular particle shapes (Bourrely et al., 1986b, 1991).

A variety of dust models have been examined in the literature for explaining the interstellar dust extinction problem. One such model considers the dust size distribution as a linear combination of several gamma functions of the form (Kocifaj, 2004):

$$f(a) = a \sum_{s=1}^{\infty} k_n \exp(-b_n a). \qquad (5.67)$$

where a is the size of the spheres. It may be seen that $f(0) = f(\infty) = 0$. This distribution, when substituted in the expression for K_{ext}, gives analytic results in the framework of the ADA. For $s = 1$, this yields:

$$K_{ext}(\lambda) = 4\pi \frac{k_1}{b_1} \frac{15h^4 + 10h^2 + 3}{(h^2 + 1)^3}, \qquad (5.68)$$

where

$$h = \frac{\lambda b_1}{4\pi(m-1)}, \qquad (5.69)$$

and $1/b_1$ is the modal radius. The parameter b_1 can be estimated by minimizing the differences between measured and calculated curves. However, it is found that although modal radius prediction is reasonably accurate, there can be significant differences between the theoretical and real extinction curves. It has been suggested that more size distributions need to be examined to get better agreement.

5.3 PLASMA DENSITY PROFILING

If the wavelength of incident radiation is much greater than the Debye length, the plasma column in the presence of an axial static magnetic field along the z-axis behaves like a dielectric rod with the dielectric tensor given by (Platzman and Ozaki, 1960):

$$(\epsilon) = \begin{pmatrix} \epsilon_1 & i\epsilon' & 0 \\ -i\epsilon' & \epsilon_1 & 0 \\ 0 & 0 & \epsilon_3 \end{pmatrix}, \qquad (5.70)$$

where

$$\epsilon_1 = \frac{1 - Q(1 + i\nu)}{(1 + i\nu)^2 - s^2}, \quad \epsilon_3 = 1 - \frac{Q}{1 + i\nu},$$

and

$$\epsilon' = \frac{Qs}{(1 + i\nu)^2 - s^2},$$

with

$$Q = (\omega_p/\omega)^2, s = (\omega_c/\omega)^2 \quad \text{and} \quad \nu = (\nu_c/\omega)^2.$$

Here, ω_p is the electron plasma frequency, ω_c is the electron cyclotron frequency, ν_c is the collision frequency, ω is the frequency of the incident wave and Debye length is defined as the scale over which mobile charge carriers screen out electric fields in plasma. Under such conditions, the Maxwell equations for TMWS reduce to the scalar equation for scattering by a homogeneous cylinder with $m^2 = \epsilon_3$. The EA is, therefore, expected to be a useful tool for analysis of the light scattered by a cylindrical plasma column (Sharma, 1986; Sharma and Dasgupta, 1987).

Ignoring ν_c/ω, in comparison with unity ϵ_3 can be approximated as:

$$\epsilon_3 = 1 - \left[\frac{n(x,y)}{n_c}\right]. \tag{5.71}$$

Here, n_c is the cut-off density which forms the upper limit beyond which radiation of wavelength λ will not propagate through the plasma and $n(x, y)$ is the plasma density. The eikonal phase shift for the scattering of electromagnetic waves from a cylindrical plasma can, therefore, be written as:

$$\chi(y)_{EA} = \frac{k}{2n_c}\int_{-\sqrt{a^2-y^2}}^{\sqrt{a^2-y^2}} n(x,y)\,dx. \tag{5.72}$$

Further, for a radially symmetric distribution, the phase shift function can be expressed as:

$$\chi(y)_{EA} = \frac{2\pi}{n_c}\int_y^0 \frac{b\,db\,n(b)}{\sqrt{b^2-y^2}}, \tag{5.73}$$

whose Abel inversion leads to:

$$kn(b) = \frac{-2n_c}{\pi}\int_b^a dy\,\frac{\chi'_{EA}}{\sqrt{y^2-b^2}}, \tag{5.74}$$

where

$$\chi'_{EA}(y) - \frac{d\chi(y)_{EA}}{dy}.$$

A simple method of obtaining $\chi(y)_{EA}$ is suggested by (3.28). The quantity $\exp[i\chi(y)_{EA}]$ is the field behind the plasma column and $\exp[i\chi(y)_{EA}] - 1$ is the added field which determines the scattered wave – see the discussion before (3.28). This added field may be obtained by using a small stop in the spatial frequency plane which just shields the focus of the beam. The amplitude $u(y)$ in the image plane may then be written as:

$$u(y) = F^{-1}\left[F\left[\exp[i\chi(y)_{EA}]\right]t(f_y)\right], \tag{5.75}$$

$$= F^{-1}\left[[F[\exp(i\chi(y)_{EA} - 1] + \delta(f_y)]t(f_y)\right], \tag{5.76}$$

where F and F^{-1} stand for Fourier and inverse Fourier transforms, f_y is the spatial frequency and $t(f_y)$ is the transmittance of the spatial filter. Assuming that only a negligible part of the spectral components of $[\exp[i\chi(y)_{EA}] - 1]$ is shielded by the stop one can write:

$$u(y) = \exp[i\chi(y)_{EA}] - 1, \tag{5.77}$$

and, hence, one can obtain the $\chi(y)_{EA}$ by scanning the intensity pattern in the image plane:

$$I(y) = |\exp[i\chi(y)_{EA}] - 1|^2 = 2[1 - \cos[\chi(y)_{EA}]]. \qquad (5.78)$$

A simple integration program then yields the electron density profile. This method is analogous to the one proposed by Brinkmeyer (1978) in the context of refractive index profiling of optical fibres.

5.4 BIOMEDICAL OPTICS

Light scattering techniques are of great interest for studying biological cells and tissues. Since most biological particles have a refractive index close to that of the surrounding medium, soft-particle approximations are ideally suited for analysis of the light scattered by biological particles. A number of workers have used the RGDA (e.g., Koch, 1968; Fiel, 1970; Wyatt, 1973) primarily for smaller cells such as bacteria. For large particles, Fraunhofer diffraction has also been employed (Fiel, 1970). Analytic formulas relating the features of scattering pattern to the geometrical and physical parameters of biological particles have been obtained in the framework of the EA and ADA too.

In conjunction with soft-particle approximations, light scattering tools have also been employed to characterize constituents in a collection of particles. For example, a biological tissue typically consists of many cells, micro-organisms, blood corpuscles, etc. which is often modeled as an ensemble of discrete particles. These scatterers may be spherical or nonspherical. Blood is an example of the biological disperse system of discoid-shaped particles. A coated particle model has also been used for analyzing light scattering by a biological cell. For tissues that have fibrous structure, a system of long cylinders is the appropriate model. Tissues – such as muscular tissues, cornea tissues and sclera tissue – belong to this class.

5.4.1 Blood optics

A large amount of work has been done on optical methods for the characterization of blood. A recent review of these methods can be found in an article by Yaroslavsky *et al.* (2002). The red blood cell (RBC) is basically a viscoelastic membrane containing a solution of hemoglobin. The unstressed RBC has a biconcave disk shape. It has good elastic properties such that it can pass through capillaries (blood vessels) with diameters less than the size of the unstressed RBC. If the elastic property of the RBC reduces, it may lead to a variety of medical problems because of insufficient delivery of oxygen to the concerned tissues. This elastic property of a RBC can be quantified by a technique known as ektacytometry. In this measurement technique a dilute RBC suspension is subjected to Couette flow between two coaxial transparent cylinders. This shearing force changes the discoid RBC to an ellipsoidal shape and the parameters of the ellipsoid can be inferred by analysis of the light scattered by the RBCs (Streekstra, 1994).

The theoretical expressions derived using the ADA show that the points of equal intensity in space constitute an elliptical curve corresponding to the ellipticity of the cell at small observation angles. If a plane wave traveling in the z-direction is incident on the elliptic cross-section of a deformed RBC, the scattering function in the framework of the ADA can be written as (Streekstra et al., 1993):

$$S(\theta) = k^2 A^2 \int_0^{\pi/2} [1 - e^{i\rho^* \sin \gamma}] J_0(kA \cos \gamma \sin \theta) \cos \gamma \sin \gamma \, d\gamma, \qquad (5.79)$$

where $A = \sqrt{ab}$ with a as the long axis and b as the short axis of the cross-sectional ellipse and:

$$\rho^* = 2kc(m - 1),$$

where c is the length of the third axis of the ellipsoid parallel to the direction of the incident wave vector. Equation (5.79) tells us that all points on the screen with the same θ (small) will be curves of equal intensity satisfying the equation of an ellipse whose ellipticity is equal to the ellipticity of the cell.

In actual measurements, a deformation index DI may be defined as:

$$DI = \frac{l - s}{l + s}$$

where l and s are the long and short axis of the scattering pattern. The DI is measured at different angular velocities of the outer cylinder. The DI is plotted against the calculated shear stress in the suspension. This represents the deformability of the RBC under consideration: the higher the stress, the higher the ellipticity. Confirmation of the validity of the relationship between the cell shape and intensity pattern as predicted by the ADA has been done by comparing ektacytometric results with true rheological measurements (Streekstra, 1994).

The relationship between cell shape and intensity pattern as predicted by the ADA is valid for a cell population too. Assuming that the RBC population follows a normal distribution, the calculations performed using the ADA show that for this case the axial ratio of the isointensity curves reflects the mean deformation of RBCs in the population. However, the relationship between the cell shape and the observed intensity pattern is limited to cases where the cells are aligned parallel to the direction of the flow and have uniform deformation. For an arbitrarily oriented ellipsoid the axial ratio of the isointensity curve represents the axial ratio of the elliptical projected area of the ellipsoid (Streekstra et al., 1994). It has been noted that at angles smaller than 2°, measurements could lead to large errors. Isointensity curves therefore need to be scanned between 2° and the first minimum of the theoretical curve.

In analyzing the scattering of electromagnetic radiation from a dilute suspension of erythrocytes, a widely used technique is Monte Carlo simulation, in which one traces the path of an incident photon. An important input required in implementing this approach is the scattering phase function which can be used to obtain the probable direction of a photon after scattering. Generally, the empirical function known as the "Henyey–Greenstein phase function" (Henyey and Greenstein, 1941) is used for this purpose. However, this does not give a very accurate representation of the exact phase function and, hence, alternative phase functions have been explored. Hammer *et al.* (1998) have investigated extinction and angle-resolved scattering from dilute suspensions of RBCs. They compared experimental measurements of phase function with predictions of Mie theory, the ADA, the RGDA and some other empirical phase functions. Measurements were in satisfactory agreement with the predictions of Mie theory. But, the best agreement found was with the anomalous diffraction phase function for small scattering angles. An approximation, similar to the ADA, known as the "Rytov approximation", has also been used to approximate the RBC phase function (He *et al.*, 2004). Formally, the Rytov approximation (Rytov, 1937) is similar to the EA. The scattering function in this approximation is expressed as:

$$S(\theta, \varphi) = -k^2 \int \exp(i\mathbf{q}.\mathbf{b}) \left[\exp\left(\sum_n R_n \right) - 1 \right] d\mathbf{b}, \qquad (5.80)$$

where

$$R_1 = i\chi_{ADA}(\mathbf{b}), \qquad (5.81)$$

and terms such as $R_2, R_3, \ldots$, etc. give higher order corrections in powers of $(n-1)$. The lowest nontrivial order of the Rytov approximation is the same as the ADA. In the work of He *et al.* (2004), the Rytov approximation is considered right down to the lowest order and the phase shift is taken to be the same as in the ADA. Numerical comparisons with exact phase functions (calculated using the FDTD and the discrete dipole approximation) for discoid-shaped RBCs were performed. It was found that the Rytov (or the ADA) gives remarkably accurate values for scattering angles up to 30°.

The light scattering properties of individual erythrocytes have also been studied by Shvalov *et al.* (1999). Angular scattering patterns in the range 10°–60° were obtained using a scanning flow cytometer. Experimental observations when contrasted with those calculated using RGDA, WKBA, two-wave WKBA and Mie calculations for a volume-equivalent sphere showed good qualitative agreement between calculated and measured phase functions. But, a large deviation occurs in the visibility, defined as:

$$V(15) = \frac{I_{max} - I_{min}}{I_{max} + I_{min}}, \qquad (5.82)$$

where I_{min} and I_{max}, respectively, denote the light intensity of the minimum and maximum that occur after 15°.

Borovoi *et al.* (1998) have developed a computer code that allows one to calculate the optical parameters of a cell, such as the absorption and scattering efficiencies and the phase function, of arbitrary shape and structure. This has been achieved within the framework of the ADA. The accuracy of the code was assessed by comparing its predictions with Mie theory results for an erythrocyte suspended in blood plasma. The difference from Mie theory calculations does not exceed 4%. This code has been used to study the dependence of optical parameters on physical parameters such as size, shape, orientations, hemoglobin concentration and the degree of oxygenation. The value of some of these parameters can be used to determine the pathology. However, no mathematical relationships between geometrical and optical parameters were obtained.

For obtaining shape and structure information Borovoi *et al.* (1998) have introduced a new concept called the "*S*-function". The *S*-function is defined as:

$$S(\mathbf{b}) = \int \Omega(\mathbf{b}')\Omega^*(\mathbf{b}' + \mathbf{b})\, d\mathbf{b}', \qquad (5.83)$$

where

$$\Omega(\mathbf{b}) = [e^{i\phi_{ADA}(\mathbf{b})} - 1]D(\mathbf{b}), \qquad (5.84)$$

with $D(\mathbf{b})$ as the shadow function, which is equal to unity inside the particle projection and is 0 outside. Thus, $\Omega(\mathbf{b})$ is nothing more than the near field behind the scatterer. The *S*-function is related to the scattering and absorption cross-section via the relation:

$$C_{sca} = \int |\Omega(\mathbf{b})|^2\, d\mathbf{b} \quad \text{and} \quad C_{abs} = \int [1 - |\Omega(\mathbf{b})|^2\, d\mathbf{b}i]. \qquad (5.85)$$

The *S*-function also determines the phase function as a two-dimensional Fourier transform of the *S*-function. Thus, it determines both cross-section and small-angle scattering pattern. Numerical comparisons show that the *S*-function is more sensitive to any change in RBC geometrical parameters and, hence, more suitable for retrieval of some RBC parameters. The *S*-function is most conspicuous for erythrocytes with some irregular inclusion. In this event, the imaginary part of the *S*-function becomes the most sensitive probe.

5.4.2 Tissue optics

Broadly speaking, a tissue may be viewed as a group of cells trapped in a network of fibres. The exact modeling of such a complicated structure is a next-to-impossible task. A simplified model used in studies relating to light propagation in a biological tissue treats tissue as a turbid medium with randomly distributed discrete scatterers. Because the nature of the problem requires *in vivo* characterization of the tissue, one is more interested in properties like diffuse reflectance and diffuse transmittance, etc. This problem may be solved using Monte Carlo simulation and – to do this – one again requires the single-scattering phase function as input which, for this problem, is simply the averaged phase function of particles of various sizes and shapes in an elementary volume where multiple scattering as well as dependent scattering are absent. Under these conditions, construction of the phase function can be achieved by simply taking a weighted superposition of the phase functions of particles of various sizes that are present in that elementary volume. The size distribution of components in the tissues varies from tissue to tissue. There is no universal distribution function. The various distribution functions that have been used are Gaussian, Gamma and power law (fractal) distributions. Soft-particle approximations can play an important role in construction of single-scattering phase functions.

The fact that forward-scattered intensity in the ADA shows oscillations with frequency proportional to $x(n-1)$ has been used by Perelman *et al.* (1998) to devise a technique for early detection of cancer. Mucosal tissues, from which precancerous changes begin, consist of a thin surface layer of epithelial cells. In normal tissues, this is a well-organized layer of cells of diameter 10–20 μm and height 25 μm. In precancerous epithelium, the cells proliferate and the cell nuclei grow as large as 20 μm in height, and can occupy almost the entire cell volume. When a beam of light is incident on this layer a portion of light is backscattered by nuclei in the cells. The remaining light is transmitted in the tissue where it undergoes multiple scattering before coming to the surface eventually, where it once again is scattered by nuclei. Thus, the emerging light consists of three components (*i*) forward scatter, (*ii*) backscatter and (*iii*) a diffuse background. Perelman *et al.* (1998) have succeeded in separating the first component from the rest and have shown that by analyzing the amplitude and frequency of the fine structure in reflectance using the ADA one can determine the size distribution and density of the nuclei, which can determine the existence of dysplasia.

5.4.3 Size and shape of bacteria

The ADA in the form of the Gaussian ray approximation (GRA) has been used to determine the size and shape of different species of bacteria from light transmission (Katz *et al.*, 2003, 2005). Let us recall that in this approximation the extinction efficiency is given by relation (3.218). In the absence of absorption, the extinction efficiency in this approximation becomes:

$$Q_{GRA}^{ext} = 2 - 2\cos[k(n-1)\mu] \exp\left[-\frac{k^2\sigma^2[(n-1)^2]}{2}\right].$$

Further, if $k(n-1)\mu < 1$, the $\cos[k(n-1)\mu]$ term and the exponential function in the above equation can be expanded. Retaining terms of the order of $k^4(n-1)^4\mu^4$, the above equation reduces to:

$$Q_{GRA}^{ext} = k^2(n-1)^2(\mu^2 - \sigma^2) - k^2(n-1)^4(\mu^4 + 6\mu^2\sigma^2 + 3\sigma^4). \quad (5.86)$$

It has been checked that the approximation is valid for soft particles in the particle size range $5 \leq x \leq 50$. This range includes most micro-organisms when visible light is involved in the measurements. The validity range of the approximation, in fact, is considerably larger for polydisperse particles. Using the GRA on randomly oriented particles, the mean geometric path $\langle l \rangle$ and mean square root geometric ray path can be determined by light extinction measurements. The nominal particle size can then be determined if the particle shape is known. Retrieval of scatterer characteristics was examined for measurements for three species of bacteria and was found to be in good agreement with scanning electron microscopy measurements.

5.4.4 Circular dichroism and optical rotation

Many biological particles exhibit circular dichroism (CD) or optical rotation (OR). CD relates to the different absorption of right-handed and left-handed circularly polarized light and OR is the rotation of the plane of polarization of linearly polarized light caused by different real parts of the refractive index for right-handed and left-handed circular polarization. Such particles are called "optically active particles"; the measurements of optical activity can be useful in studies of bio-particles (see, e.g., Bohren and Huffman, 1983; Rosen and Pendleton, 1995; Kokhanovsky, 2005).

For an optically thin medium consisting of dispersed particles, CD and OR can be expressed as (Bohren and Huffman, 1983):

$$OR = \frac{\pi N}{k^2} \text{Re}(i\Delta S), \quad CD = \frac{\pi N}{k^2} \text{Im}(i\Delta S), \quad (5.87)$$

where

$$\Delta S = S_L(0) - S_R(0). \quad (5.88)$$

The subscripts L and R correspond to left-handed and right-handed circular polarizations of the incident wave, respectively, and N is the number of particles per unit volume.

Equations (5.87) make it possible to study $OR(\lambda)$ and $CD(\lambda)$ if $\Delta S(0)$ can be calculated. This may be done using exact solutions, whenever available. Approximation methods may be used when either exact solutions are not available or if the exact solution is so complex that approximate solutions are preferable. The ADA has been used for this purpose (see, e.g., Kokhanovsky, 2005). For a polydispersion characterized by the particle size distribution $f(a)$ and $|m_L - m_R| \ll 1$, a lengthy but straightforward calculation shows that:

$$\begin{pmatrix} OR \\ CD \end{pmatrix} = \begin{pmatrix} D'' & D' \\ D' & -D'' \end{pmatrix} \begin{pmatrix} \pi \Delta n c/\lambda \\ \pi \Delta n' c/\lambda \end{pmatrix}, \quad (5.89)$$

where $D = D' + iD''$, $\Delta m = m_L - m_R = \Delta n + i\Delta n'$ and:

$$D(\rho^*) = \frac{\int_0^\infty a^3 f(a) \, d(\rho^*) \, da}{\int_0^\infty a^3 f(a) \, da}, \quad (5.90)$$

with $d(\rho^*) = 3(\partial \mathcal{K}/\partial \rho^*)_{\rho^* = \bar{\rho}^*}$. Equation (5.89) tells us that the dependence of OR and CD on particle dimensions is through the functions D' and D''. For nonabsorbing spherical particles these can be calculated analytically for a number of particle size distributions. For a monodispersion of nonabsorbing spherical particles D' and D'' can be calculated to be:

$$D'(\rho) = \frac{6(\cos\rho - 1)}{\rho^3} + \frac{6\sin\rho}{\rho^2} - \frac{3\cos\rho}{\rho}, \quad (5.91)$$

and

$$D''(\rho) = \frac{3\sin\rho}{\rho} + \frac{6\cos\rho}{\rho^2} - \frac{6\sin\rho}{\rho^3}. \quad (5.92)$$

Here, $\rho = x(m_R + m_L - 2)$. It follows that the smallest value of rotation of the polarization plane corresponds to the maximum value of CD and vice versa. It may be noted that equations (5.91) and (5.92) are identical to equations (5.60) and (5.59), respectively. Hence, the zeros of the functions $D'(\rho)$ and $D''(\rho)$ approximately coincide with the maxima of the functions $\mathcal{B}'(\rho)$ and $\mathcal{B}''(\rho)$.

5.5 OCEAN OPTICS

The natural waters in sea and other large water bodies generally consist of water molecules and dissolved and suspended impurities. Computability of the optical properties of particles suspended in ocean waters is important because these serve as input in the radiative transfer calculations in ocean waters and, hence, in the energy balance calculations in the environment. Soft-particle approximations – in particular, the ADA – has played an important role in such computations. It has been recognized that it can adequately describe many of the observed quantities – such as attenuation, absorption and total scattering of algal cells – modelled as homogeneous spheres (Bricaud and Morel, 1986). However, the homogeneous sphere model has been inadequate for reproducing angular scattering data (Quinby-Hunt et al., 1989; Volten et al., 1998; Vaillancourt et al., 2004). For this reason, the phase function used in the solution of the radiative transfer equation is an empirical phase function (Haltrin, 2002).

Coated sphere models have been adapted for some species with chloroplast as the core by Quinby-Hunt et al. (1989). Quirantes and Bernard (2004) have examined a nonconcentric coated sphere model as well as coated spheroids (centered). Comparison of the three models – concentric sphere model, nonconcentric two-layer model and concentric spheroid model – shows that the scattering and extinction efficiencies in all three models are very similar. The absorption efficiency factor is also found to be nearly shape-independent. Similar results have been obtained for inhomogeneous scatterers by Aas (1984), Kitchen and Zaneveld (1992), Zaneveld and Kitchen (1995) and Bricaud et al. (1992). Backscattering efficiency is found to be quite sensitive to shape.

In a simple model, phytoplankton can be regarded as made up of two main constituents – namely, (*i*) dry matter and (*ii*) water. Three extreme ways in which these constituents may be organized are:

(*a*) a homogeneous sphere consisting of water and dry matter mixed perfectly;
(*b*) a coated sphere with dry matter as the hard shell filled with water; and
(*c*) a coated sphere with dry matter as the hard core.

An investigation into the influence of internal structure on the predictions of the ADA was done by Aas (1984). The following conclusions were reached:

1. For small particles there is hardly any dependence on structure. Scattering, on the other hand, does depend on the internal structure. It was noted that scattering is greater when the scattering and absorbing material is concentrated towards the center.
2. For large particles, scattering and absorption are smaller when the refracting and absorbing material is concentrated towards the center of the particle.
3. For large particles, the concentration of material towards the outer surface has no effect on absorption and a much smaller effect on scattering.

A formula for absorption efficiency – developed from (4.137) – reads as (see, e.g., Paramonov, 1996):

$$C_{abs} = [1 - \exp(-2\nu/3)]\langle S \rangle, \qquad (5.93)$$

where

$$\nu = \frac{3}{2}\frac{\langle V \rangle \tilde{\alpha}}{\langle S \rangle}, \qquad (5.94)$$

where $\tilde{\alpha} = 4\pi n'/\lambda$ and $\langle V \rangle$ and $\langle S \rangle$ are the volume and cross-section, respectively, averaged over an ensemble. Paramonov (1994, 1995) has confirmed theoretically the validity of this formula for randomly oriented, soft ellipsoidal particles.

5.6 MISCELLANEOUS APPLICATIONS

The potential function $U(\mathbf{r})$ for a system of N spheres of a_i and refractive indices m_i, centered at $\mathbf{R}_i$, can be expressed as:

$$U(\mathbf{r}) = -k^2 \sum_i (m_i^2 - 1)\Theta(a_i - |\mathbf{r} - R_i|)$$

The EA with this potential function has been tested numerically and is found to work fairly well when the gaps between spheres is small (Chen, 1990). A particular case where the EA greatly simplifies the problem is the case of closely packed dielectric spheres of the same size but not necessarily the same refractive index, lined up along a common axis. It can easily be verified that in this case the scattering function is equal to the scattering function of a single sphere of refractive index $m_{eff} = 1 + \sum(m_i - 1)$. This gives the sum rule:

$$S(q, a, m_1 \ldots m_N) = S(q, a, n_{eff}).$$

The results of the sum rule have been found to be reasonably good (Chen, 1990).

Optical bistability is characterized by the different light transmission states of an optical system for a given light intensity input. In practice, a nonlinear absorbing medium in an optical resonator constitutes an optical bistable device. The Fabry–Perot resonator (FPR) with a dispersively nonlinear medium is an example of such a device. For this configuration, the exact solution of Maxwell's equations has been obtained by Marburger and Felber (1978). For more complicated configurations, however, exact solutions are not available. Formulation and solution of the problem then requires use of some approximate methods. Orenstein et al. (1984, 1985, 1986, 1987a, b) have shown that the problem can be formulated for arbitrary configurations in a straightforward way in the EA and that solutions can be obtained numerically. In some cases, analytic solutions could also be obtained. The solutions obtained in the EA confirm the multistability of the output intensity. It may be pointed out here that what has been referred to as the EA in the work of Orenstein and collaborators is, in effect, little more than the ADA in the language of this book.

A typical cladded optical fiber is characterized by a dielectric constant (Calvo and Juncos del Egido, 1979):

$$m^2(|\mathbf{b}|) = m_2^2 \quad \text{for } a_2 \geq |\mathbf{b}| \geq a_1 \tag{5.97}$$

and

$$m^2(\mathbf{b}) = m_1^2 \left[1 - 2\delta \left(\frac{|\mathbf{b}|}{a_1}\right)^2 + \Delta \delta^2 \left(\frac{|\mathbf{b}|}{a_1}\right)^4 \right] > m_2 \quad \text{for } a_1 \geq |\mathbf{b}| \geq 0, \tag{5.98}$$

where a_1 is the radius of the core, a_2 is the radius of the cladding, and δ and Δ are limited to values between 0 and 1. For the following typical values of the cladded fiber $a_1 = 2\,\mu\text{m}$, $a_2 = 50\,\mu\text{m}$, $m_1 = 1.52$, $m_2 = 1.50$, $\delta = \Delta = 10^{-2}$, the conditions for applicability of the EA are adequately satisfied. Scattered intensities can be readily arrived at by using the formulas of Section 3.5. For this index profile, it has been noted that the contribution of the core to the diffraction pattern is small. It is the cladding that dominates the diffraction pattern. Also, the effect of the quartic term is found to be almost negligible: that is, the diffraction pattern obtained with quadratic power is almost similar to that obtained using a quartic index profile.

Calvo and Juncos del Egido (1982) have considered the diffraction of electromagnetic waves by a volume hologram in the framework of the EA. The problem is analogous to that of the diffraction of light by ultrasonic waves. Indeed, the solutions obtained by Calvo and Juncos del Egido agree with those obtained by Raman and Nath (1935, 1936) in the context of scattering by ultrasound waves. Calvo and Juncos del Egido have also examined corrections made to the EA. A numerical evaluation of these corrections to several orders of diffraction has been carried out and the following observations regarding the importance of corrections with order of diffraction have been made: (*i*) the correction to zero-order vanishes exactly; (*ii*) the most important correction occurs for the first order of diffraction; (*iii*) at higher orders the importance of the correction decreases as the order of diffraction increases.

Appendix A

Scattering formulas in the anomalous diffraction approximation for an arbitrarily oriented hexagonal column

The following analytic expressions have been obtained for the forward scattering function by Sun and Fu (1999):

$$S(0)_{ADA} = \frac{k^2}{2\pi}[2(A_1 + A_2) + A_3] \quad \text{for } 0 \leq \beta \leq \arctan\left[\frac{l}{\sqrt{3}a/\cos\alpha}\right], \quad \text{(A.1)}$$

$$S(0)_{ADA} = \frac{k^2}{2\pi}[2A_1 + B_3 + 2(a_2 + b_2)]$$

$$\text{for } \arctan\left[\frac{l}{\sqrt{3}a/\cos\alpha}\right] \geq \beta \geq \arctan\left[\frac{l}{(\sqrt{3}/2)a/\sin(\pi/6+\alpha)}\right], \quad \text{(A.2)}$$

and

$$S(0)_{ADA} = \frac{k^2}{2\pi}[2B_2 + B_3 + 2(a_1 + b_1)] \quad \text{for } \arctan\left[\frac{l}{(\sqrt{3}/2)a/\sin(\pi/6+\alpha)}\right] \leq \alpha, \quad \text{(A.3)}$$

where l is the length of a hexagonal column, α is the azimuth angle which because of hexagonal symmetry varies from 0 to $\pi/6$, and β is the elevation angle which varies from 0 to $\pi/2$ ($\beta = 0°$ corresponds to perpendicular incidence). The y-axis is perpendicular to one of the side planes of the hexagon. The quantities occurring in

(A.1)–(A.3) are as follows:

$$A_1 = Aa\sin(\pi/6 - \alpha) + \frac{\sqrt{3}}{4}a^2 \sin\beta \frac{\sin(\pi/6 - \alpha)}{\sin(\pi/6 + \alpha)}$$

$$+ AE\left(1 - e^{(\sqrt{3}aB/2\sin(\pi/6+\alpha))}\right)$$

$$+ \frac{\sin\beta}{B}\left[a\sin(\pi/6 - \alpha)e^{(\sqrt{3}aB/2\sin(\pi/6+\alpha))} + E(1 - e^{(\sqrt{3}aB/2\sin(\pi/6+\alpha))})\right], \quad (A.4)$$

$$A_2 = (A + M\sin\beta)\sqrt{3}a\sin\alpha + 1.5Na^2 \sin\beta \sin^2\alpha$$

$$+ \frac{e^{BM}}{BN}(A - M\sin\beta)\left(1 - e^{BN\sqrt{3}a\sin\alpha}\right) + G, \quad (A.5)$$

$$A_3 = \left[A + \sqrt{3}a\frac{\sin\beta}{\cos\alpha} - \left(A - \sqrt{3}a\frac{\sin\beta}{\cos\alpha}\right)e^{B\sqrt{3}a/\cos\alpha}\right]a(\cos\alpha - \sqrt{3}\sin\alpha), \quad (A.6)$$

$$B_2 = (C + DM)\sqrt{3}a\sin\alpha + 1.5DNa^2 \sin^2\alpha, \quad (A.7)$$

$$B_3 = (C\cos\alpha + \sqrt{3}aD)a(1 - \sqrt{3}\tan\alpha), \quad (A.8)$$

$$a_1 = AJ + \frac{\sqrt{3}}{3}l^2\frac{\cos\beta}{\tan\beta}\sin(\pi/6 - \alpha)\sin(\pi/6 + \alpha) + AE(1 - e^{Bl\cot\beta})$$

$$\times \frac{\sin\beta}{B}\left[Je^{Bl\cot\beta} + E(1 - e^{Bl\cot\beta})\right], \quad (A.9)$$

$$b_1 = C[a\sin(\pi/6 - \alpha) - J] + \frac{\sqrt{3}}{4}Da^2\frac{\sin(\pi/6 - \alpha)}{\sin(\pi/6 + \alpha)}\left[1 - \frac{4l^2\sin^2(\pi/6 + \alpha)}{3a^2\tan^2\beta}\right], \quad (A.10)$$

$$a_2 = (A + M\sin\beta)(\sqrt{3}a\sin\alpha - I) + 0.5N\sin\beta(3a^2\sin^2\alpha - I^2)$$

$$+ (A - M\sin\beta)\frac{e^{BM}}{BN}(e^{BNI} - e^{BN\sqrt{3}a\sin\alpha}) + F, \quad (A.11)$$

$$b_2 = CI + DMI + 0.5DNI^2, \quad (A.12)$$

where I, J, E, F, M, N and G are given by:

$$I = -a\cos\alpha + \frac{2l\cos\alpha \sin(\pi/6 + \alpha)}{\sqrt{3}\tan\beta}, \quad (A.13)$$

$$J = \frac{2l\sin(\pi/6 - \alpha)\sin(\pi/6 + \alpha)}{\sqrt{3}\tan\beta}, \quad (A.14)$$

$$E = \frac{2\sin(\pi/6 - \alpha)\sin(\pi/6 + \alpha)}{\sqrt{3}B}, \quad (A.15)$$

$$F = \frac{\sin\beta e^{BM}}{B}\left[\left(\sqrt{3}a\sin\alpha - \frac{1}{BN}\right)e^{BN\sqrt{3}a\sin\alpha} - \left(I - \frac{1}{BN}\right)e^{BNI}\right], \quad (A.16)$$

$$M = \frac{\sqrt{3}a}{\cos \alpha},\tag{A.17}$$

$$N = -\frac{\sqrt{3}}{2\cos \alpha \sin(\pi/6 + \alpha)},\tag{A.18}$$

$$G = \sin \beta \frac{E^{BM}}{B}\left[\left(\sqrt{3}a\sin \alpha - \frac{1}{BN}\right)e^{BN\sqrt{3}a\sin \alpha} + \frac{1}{BN}\right].\tag{A.19}$$

The extinction cross-section is given by $C_{ext} = (4\pi/k^2)S(0)$. The absorption cross-section C_{abs} can be obtained by replacing $(2\pi/k^2)S(0)$ by C_{abs} and also replacing the A, B, C and D in the previous formulas by the following expressions:

$$A = l\cos \beta - \frac{\sin \beta \cos \beta}{kn'},\tag{A.20}$$

$$B = -\frac{2kn'}{\cos \beta},\tag{A.21}$$

$$C = 2l\cos \beta - \left(l\cos \beta + \frac{\sin \beta \cos \beta}{kn'}\right)(1 - e^{-2kln'/\sin \beta}),\tag{A.22}$$

$$D = \sin \beta (1 - e^{-2kln'/\sin \beta}).\tag{A.23}$$

To calculate the extinction and absorption efficiency factor one requires the projected area of the particle on which radiation is incident. This can be shown to be:

$$p = \frac{3\sqrt{3}}{2}a^2 \sin \beta + 2al \cos \beta \cos \alpha.\tag{A.24}$$

The extinction and absorption efficiency factor are then given by $Q_{ext} = C_{ext}/p$ and $Q_{abs} = C_{abs}/p$.

Appendix B

Addition theorems employed in deriving the main form of the Perelman approximation (MPA) for a spherical particle

The addition theorems for two Bessel functions are well-known and can be written as (see, e.g., Perelman, 1991; Gradshteyn and Ryzhik, 1980):

$$\sum_{n=0}^{n=\infty}(2n+1)\psi_n(x_i)\psi_n(x_j)P_n(\mu) = x_i x_j s(\omega_{ij}), \tag{B.1}$$

$$\sum_{n=0}^{n=\infty}(2n+1)\psi_n(x_i)\chi_n(x_j)P_n(\mu) = x_i x_j \sigma(\omega_{ij}), \quad |x_i| < x_j \tag{B.2}$$

where

$$\omega_{ij} \equiv \omega_{ij}(\mu) = (x_i^2 + x_j^2 - 2x_i x_j \mu)^{1/2},$$

Using relations (B.1) and (B.2) one can also obtain the following relations for the product of four Bessel functions (Perelman, 1991):

$$\sum_{n=0}^{n=\infty}(n+0.5)\prod_{r=1}^{r=4}\frac{\psi_n(x_r)}{x_r} = \frac{1}{4}\int_{-1}^{1}\frac{\sin\omega_{12}\sin\omega_{34}}{\omega_{12}\omega_{34}}d\mu, \tag{B.3}$$

$$\sum_{n=0}^{n=\infty}(n+0.5)\prod_{r=1}^{r=3}\frac{\psi_n(x_r)}{x_r}\frac{\chi_n(x_4)}{x_4} = \frac{1}{4}\int_{-1}^{1}\frac{\sin\omega_{12}\cos\omega_{34}}{\omega_{12}\omega_{34}}d\mu. \tag{B.4}$$

Deduction of the addition theorems that involve the derivatives of the Bessel and Neumann functions is somewhat more involved. However, Perelman (1991) has successfully obtained addition theorems for such products too. It can be verified that using relations (B.3) and (B.4), one can obtain the following relations:

$$\sum_{n=0}^{\infty}(n+0.5)\psi_n'^2(y)\psi_n^2(x) = \frac{1}{128m}\bigg[(4+2R\rho)[\text{ci}(R)-\text{ci}(\rho)]$$
$$+R^2-\rho^2+(\rho^2+R^2+4R\rho)[s(R)-s(\rho)]$$
$$+(3R^2+4R\rho-\rho^2)c(R)-(3\rho^2+4R\rho-R^2)c(\rho)\bigg], \quad (B.5)$$

$$\sum_{n=0}^{\infty}(n+0.5)\psi_n'(y)\psi_n(y)\psi_n'(x)\psi_n(x) = \frac{1}{128}\bigg[4[\text{ci}(R)-\text{ci}(\rho)]$$
$$+(R^2-\rho^2)[S(R)+S(\rho)]$$
$$+(3R^2+\rho^2)c(R)-(3\rho^2+R^2)c(\rho)\bigg], \quad (B.6)$$

$$\sum_{n=0}^{\infty}(n+0.5)\psi_n'^2(y)\psi_n(x)\chi_n(x) = \frac{1}{128m}\bigg[(4+2R\rho)[\text{si}(R)-\text{si}(\rho)]$$
$$+(R^2+\rho^2+4R\rho)[\sigma(R)-\sigma(\rho)]$$
$$+(3R^2+4R\rho-\rho^2)\gamma(R)+(3\rho^2+4R\rho-R^2)\gamma(\rho)\bigg], \quad (B.7)$$

$$\sum_{n=0}^{\infty}(n+0.5)\psi_n^2(y)\psi_n'(x)\psi_n'(y) = \frac{m}{128}\bigg[(4-2R\rho)[\text{si}(R)-\text{si}(\rho)]2$$
$$+(\rho^2+R^2-4R\rho)[\sigma(R)-\sigma(\rho)]$$
$$+(3R^2-4R\rho-\rho^2)\gamma(R)$$
$$+(3\rho^2-4R\rho-R^2)\gamma(\rho)\bigg], \quad (B.8)$$

and

$$\sum_{n=0}^{\infty}(n+0.5)\psi_n(y)\psi_n(y)[\psi_n'(x)\chi_n(x)+\psi_n(x)\chi_n'(x)]$$
$$= \frac{1}{64}\bigg[4[\text{si}(R)-\text{si}(\rho)]+(-\rho^2+R^2)[\sigma(R)-\sigma(\rho)]$$
$$+(3R^2+\rho^2)\gamma(R)+(3\rho^2+R^2)\gamma(\rho)\bigg], \quad (B.9)$$

where
$$c(z) = \frac{(1-\cos z)}{z^2}, \quad \gamma(z) = \frac{\sin(z)}{z},$$
$$\operatorname{ci}(z) = \int_0^z \frac{(1-\cos t)\,dt}{t} \quad \text{and} \quad \operatorname{si}(z) = \int_0^z \frac{\sin t\,dt}{t},$$
$$s(z) = \frac{\sin(z)}{z}, \quad \sigma(z) = \frac{\cos(z)}{z},$$

and
$$\rho = 2x(m-1), \quad \mu = \cos\theta, \quad R = 2x(m+1).$$

Appendix C

Derivation of Perelman approximation for the light scattered by an infinitely long cylinder

We begin by recalling that, in the limit $m \to 1$, terms in the denominators of scattering coefficients a_l and b_l for an infinitely long circular cylinder may be approximated as (Section 4.2.5):

$$h_{1n} = h_{2n} = 0; \quad h_{3n} = h_{4n} = -\frac{2m^n}{\pi x} = -\frac{2}{\pi x}. \tag{C.1}$$

In arriving at (C.1), terms of order $(m-1)$ or of higher order in $(m-1)$ have been ignored and standard relations (Gradshteyn and Ryzhik, 1980):

$$(J_{n+1}(x)N_n(x) - J_n(x)N_{n+1}(x)) = \frac{2}{\pi x}, \tag{C.2}$$

and

$$\zeta_n(x) = m^n \sum_{k=0}^{k=\infty} \frac{(-1)^k (m^2-1)^k}{k!} \zeta_{n+k}(x); \quad |m^2 - 1| \ll 1, \tag{C.3}$$

have been used. In equation (C.3), ζ_n could be the Bessel function J_n or the Neumann function N_n. The scattering function for transverse magnetic wave scattering:

$$T(\theta)_{TMWS} = \frac{i\pi x}{2} \sum_{n=-\infty}^{\infty} b_n e^{-in\theta}, \tag{C.4}$$

where

$$b_n = h_{1n}/(h_{1n} + ih_{3n}),$$

may then be written, with the help of (C.1), as:

$$T(\theta)_{PA}^{TMWS} = \frac{i\pi x}{2} \sum_{n=-\infty}^{\infty} h_{1n} e^{-in\theta}, \tag{C.5}$$

where

$$h_{1n} = mJ_n'(mx)J_n(x) - J_n(mx)J_n'(x).$$

Then, by virtue of the relation (Gradshteyn and Ryzhik, 1980):

$$x[m\zeta_n(x)J_n'(mx) - J_n(mx)\zeta_n'(x)] = (1 - m^2) \int^x \alpha J_n(m\alpha)\zeta_n(\alpha)\, d\alpha \qquad (C.6)$$

one can write:

$$T(\theta)_{PA}^{TMWS} = -\frac{i\pi}{2}(m^2 - 1) \int^x \alpha \sum_{n=-\infty}^{\infty} J_n(m\alpha)J_n(\alpha)\, e^{-in\theta}\, d\alpha. \qquad (C.7)$$

The summation in (C.7) can be performed with the help of the relation (Gradshteyn and Ryzhik, 1980):

$$\zeta_n(\alpha v) = \sum_{n=-\infty}^{\infty} J_n(m\alpha)\zeta_n(\alpha)\, e^{-in\theta}, \qquad (C.8)$$

to yield:

$$T(\theta)_{PA}^{TMWS} = -\frac{i\pi}{2}(m^2 - 1) \int^x \alpha J_0(\alpha w(\cos\theta))\, d\alpha$$

$$= -\frac{i\pi x}{2}(m^2 - 1) \frac{J_1(x w(\cos\theta))}{w(\cos\theta)}, \qquad (C.9)$$

where $w(t) = (1 + m^2 - 2mt)^{1/2}$. Equation (C.9) is nothing more than the expression given in the text.

To obtain the main form of the Perelman approximation (MPA), it is convenient to express the scattering coefficient b_n as:

$$b_n = \frac{h_{1n}(h_{1n} - ih_{3n})}{h_{1n}^2 + h_{3n}^2},$$

or

$$b_n = \frac{h_{1n}(mJ_n'(mx)H_n(x) - J_n(mx)H_n'(x))}{h_{1n}^2 + h_{3n}^2}, \qquad (C.10)$$

where $H_n(x) = J_n(x) - iN_n(x)$ is the Hankel function of the second kind. Then, using (C.1), (C.5) and (C.9) it can easily be shown that:

$$T(\theta)_{MPA}^{TMWS} = \frac{\pi^2 x^2 (m^2 - 1)^2}{4} \int^x \alpha\, d\alpha \int^x \beta\, d\beta \sum_{n=-\infty}^{\infty} J_n(m\alpha)J_n(\alpha)J_n(m\beta)H_n(\beta)\, e^{-in\theta}. \qquad (C.11)$$

The sum in (C.11) can be expressed as:

$$\sum_{n=-\infty}^{\infty} J_n(m\alpha)J_n(\alpha)J_n(m\beta)H_n(\beta)\, e^{-in\theta} = \frac{1}{2\pi} \int_0^{2\pi} \sum_n J_n(m\alpha)J_n(\alpha)\, e^{-in(\theta+\phi)}$$

$$\times \sum_p J_p(m\beta)H_p(\beta)\, e^{+ip\phi} \qquad (C.12)$$

with the help of a delta function. This allows the use of standard addition theorems (C.8) for two Bessel–Hankel functions for sums over n and p separately and one

obtains:
$$T(\theta)_{MPA}^{TMWS} = \pi(m^2 - 1)^2 \int^x \alpha\, d\alpha \int^x \beta\, d\beta \int_0^{2\pi} d\phi J_0(\omega(\cos\theta)\alpha) H_0(\omega(\cos(\phi - \theta))\beta). \tag{C.13}$$

The integrations over α and β can be performed analytically to yield the desired result.

Appendix D

Mean value theorem and estimation of the key parameters of the distribution

For equation (5.22) to reflect a real situation, the lower and the upper limits of the integral should be taken to be a_0 and a_m. Thus, we have:

$$K_{ext} = 2\pi N \int_{a_0}^{a_m} Q_{ext}(m, ka) a^2 f(a) \, da, \qquad (D.1)$$

where the distribution function $f(a)$ obeys the relation:

$$\int_{a_0}^{a_m} f(a) \, da = 1. \qquad (D.2)$$

Starting with (D.1), it can be shown that expressions for the key parameters of the distribution that characterize $f(a)$ can be obtained by using the first mean value theorem of integral calculus (Roy and Sharma, 1997). The mean value theorem states that if $f(x)$ and $g(x)$ are two functions that are both bounded and R-integrable in $[a, b]$, and if $g(x)$ has the same sign in $[a, b]$, then:

$$\int_a^b f(x) g(x) \, dx = M \int_a^b g(x) \, dx; \quad M_1 \leq M \leq M_2,$$

where M_1 and M_2 are the infimum and supremum, respectively, of $f(x)$ in $[a, b]$. If, in particular, $f(x)$ is continuous in $[a, b]$, then:

$$\int_a^b f(x) g(x) \, dx = f(\tilde{x}) \int_a^b g(x) \, dx,$$

where the point $\tilde{x}$ is contained in $[a, b]$.

If $K_{ext}(k)$ shows a maximum at $k = k_m$, then:

$$K_{ext}(k_m) = 2\pi N \int_{a_0}^{a_m} Q_{ext}(k_m a) a^2 f(a) \, da. \qquad (D.3)$$

Appendix D

Because $Q_{ext}(k_m a)a^2$ as well as $f(a)$ are positive-valued, bounded, R-integrable and continuous in the interval $[a_0, a_m]$, we may apply the first mean value theorem of the integral calculus to (D.3) and write:

$$K_{ext}(k_m) = 2\pi N f(\tilde{a}) \int_{a_0}^{a_m} Q_{ext}(k_m a) a^2 \, da, \tag{D.4}$$

where $a_0 \leq \tilde{a} \leq a_m$. Here, we have assumed continuity of $f(a)$ in $[a_0, a_m]$. Near k_m, $K_{ext}(k)$ is stationary. Hence, it is reasonable to write:

$$\left[\frac{\partial K_{ext}(k)}{\partial k}\right]_{k=k_m} = 2\pi N f(\tilde{a}) \int_{a_0}^{a_m} \left[\frac{\partial Q_{ext}(ka)}{\partial k}\right]_{k=k_m} a^2 \, da = 0, \tag{D.5}$$

which may also be re-written as:

$$\Phi(a_m) - \Phi(a_0) = 0, \tag{D.6}$$

where

$$\Phi(\alpha) = \int_0^\alpha \left[\frac{\partial Q_{ext}(ka)}{\partial k}\right]_{k=k_m} a^2 \, da = 0, \tag{D.7}$$

and $\alpha = a_0$ or a_m. If a_0 is known a priori, we can easily obtain a_m with the help of (D.6) and (D.7). Otherwise, because a_0 is a solution of $\Phi(a_0) = 0$, we can obtain the value of a_m by solving the equation $\Phi(a_m) = 0$. Although the supposition $a_0 = 0$ is unphysical, it has been checked that the a_m determined in this way is not sensitive to the choice of a_0 if indeed a_0 is small. Expressions for $\Phi(a_m)$ and $\Phi(a_0)$ take a simple form in the anomalous diffraction approximation (ADA). In this approximation $\Phi(\alpha)$ has the analytic expression:

$$\Phi(\alpha) = \frac{1}{(m-1)^3 k_m^4} \left[\rho_m + \frac{\rho_m^2}{2} \sin \rho_m + 2\rho_m \cos \rho_m - 3 \sin \rho_m\right], \tag{D.8}$$

where $\rho_m = 2k_m \alpha (m-1)$.

Next, if we consider the region where the $K_{ext}(k)$ has an approximately linear growth in k, then differentiating (D.1) with respect to k, we may write:

$$\frac{\partial K_{ext}}{\partial k} = 2\pi N \int_{a_0}^{a_m} \frac{\partial Q_{ext}}{\partial (ka)} a^3 f(a) \, da. \tag{D.9}$$

It can be shown that in the range $ka[= R(ka)]$ for which $\partial Q_{ext}/\partial(ka)$ is positive, the mean value theorem applied to (D.9) gives for any $\tilde{k}$ such that $\tilde{k}a \in R(ka)$ for all $a_0 \leq a \leq a_m$:

$$\left.\frac{\partial K_{ext}}{\partial k}\right|_{k=\tilde{k}} = 2\pi N f(\tilde{a}) \int_{a_0}^{a_m} \left.\frac{\partial Q_{ext}}{\partial (ka)}\right|_{k=\tilde{k}'} a^3 f(a) \, da, \tag{D.10}$$

for $a_0 \geq \tilde{a}' \geq a_m$. If k_s happens to be a stationary point for the function $\partial K_{ext}/\partial k$, then we can write:

$$\int_{a_0}^{a_m} \left.\frac{\partial^2 Q_{ext}(ka)}{\partial k^2}\right|_{k=k_s} a^2 \, da = 0. \tag{D.11}$$

Now we may obtain a_m by either assuming a_0 to be 0 or assigning to it an *a priori* known value. Employing ADA we can write:

$$\Psi(a_m) - \Psi(a_0) = 0, \tag{D.12}$$

where

$$\Psi(\alpha) = \frac{1}{(m-1)^3 k_m^5}\left[3\rho_s + \frac{\rho_s^3}{2}\cos\rho_s + 3\rho_s^2 \sin\rho_s - 9\cos\rho_s - 12\sin\rho_s\right]. \tag{D.13}$$

In (D.13), $\rho_s = 2(m-1)k_s\alpha$ and α is either a_m or a_0. In practice, one may have to resort to large-scale backward extrapolation of the experimental data to have access to the true first linear stretch in the graph of $K_{ext}(k)$ versus k. As equations (D.6) and (D.12) are transcendental in nature, each may yield more than one value of a_m. Hence, the value of a_m that simultaneously satisfies (D.6) and (D.12) is to be taken.

D.1 ESTIMATION OF THE PARTICLE NUMBER N

For this purpose, let us write $Q_{ext}(ka)a^2 = G(ka)$. Clearly $G(ka)$ is positive-valued, R-integrable and continuous in the interval $[a_0, a_m]$. Hence, application of the first mean value theorem in equation (D.1) yields:

$$K_{ext} = 2\pi N G(k\xi), \quad a_0 \leq \xi \leq a_m. \tag{D.14}$$

Equation (D.14) is purely formal in nature in the sense that the mean value theorem does not say how to determine the value of ξ. Although ξ will be a function of k, it is assumed that ξ remains constant over a small range of k values. If, in the $K_{ext}(k)$-versus-k graph, we consider k_m and two adjacent points on either side of it, which we denote as k_m^- and k_m^+, then to match K_{ext} at k_m on the right-hand side of (D.14), the value of ξ should be chosen (call it $\xi^{(k_m)}$) such that the following conditions are met:

1. $G(k_m a) > G(k - m^+ a)$; $G(k_m a) > G(k_m^- a)$ for $a = \xi^{(k_m)}$.
2. $G(k_m^+ a) = G(k_m a)$ as $k_m^+ \to k_m$ and $G(k_m^- a) = G(k_m a)$ as $k_m^- \to k_m$ for $a = \xi^{(k_m)}$.
3. For the choice of any other pair (k_m^-, k_m^+) such that $K_{ext}(k_m^-) = K_{ext}(k_m^+)$, the value of ξ should remain stable.

Having chosen $\xi^{(k_m)}$ in the above manner, it has been ensured that near $k = k_m$, equation (D.14) is meaningful and, hence, for $k = k_m$ we have:

$$N = \frac{K_{ext}(k_m)}{2\pi G(k_m \xi^{(k_m)})}.$$

D.2 ESTIMATION OF THE FIRST MOMENT OF $f(a)$

For this purpose, we write equation (D.1) as:

$$K_{ext} = 2\pi N \int_{a_0}^{a_m} H(ka)\zeta(a)\, da, \qquad (D.15)$$

where $aQ_{ext}(ka) = H(ka)$ and $af(a) = \zeta(a)$. Both these functions are positive-valued, bounded, R-integrable and continuous in the interval $[a_0, a_m]$. Applying the mean value theorem one gets:

$$K_{ext} = 2\pi N H(k\mu) \int_{a_0}^{a_m} af(a)\, da = 2\pi N H(k\mu)\bar{a}, \qquad (D.16)$$

where $a_0 \leq \mu \leq a_m$. Again proceeding in the same manner as done before, one can write:

$$\bar{a} = \frac{K_{ext}(k_m)}{2\pi N H(k_m \mu^{(k_m)})}.$$

With similar reasoning one can also obtain higher order moments of the distribution $f(a)$.

Appendix E

Pearson method

If reliable data for the first three central moments μ_1, μ_2 and μ_3 are available, then one can obtain the expression for the distribution function from the following differential equation (Elderton and Johnson, 1969):

$$\frac{1}{f(a)} \frac{df(a)}{da} = -\frac{a + \frac{\mu_3}{2\mu_2}}{\mu_2 + \frac{\mu_3}{2\mu_2} a}. \tag{E.1}$$

Similar expressions exist for cases when reliable moments up to second, fourth, etc. orders are available. Equation (E.1) may be written as:

$$\frac{1}{f(a)} \frac{df(a)}{da} = -\frac{a + p}{ap + q}, \tag{E.2}$$

where $q = \mu_2$ and $p = \mu_3/2\mu_2$. A straightforward integration then yields:

$$\int \frac{df(a)}{f(a)} = -\frac{a}{p} + \left(\frac{q}{p^2} - 1\right) \ln(ap + q) + \ln C. \tag{E.3}$$

which finally gives:

$$f(a) = C \exp(-a/p)(ap + q)^{(q/p^2 - 1)}, \tag{E.4}$$

where C is a suitable normalization constant chosen such that $f(\text{mode}) = 1$.

References

Aas, E. (1984). *Some Aspects of Light Scattering by Marine Particles*. Institute of Geophysics, Oslo.
Abarbanel, H. D. I. (1972). Eikonal expansion in non-relativistic and relativistic scattering theory. In: D. Bessis (ed.), *Cargese Lectures in Physics* (Vol. 5, pp. 519–547). Gordon & Breach, New York.
Abarbanel, H. D. I. and Itzykson, C. (1969). Relativistic eikonal expansion. *Phys. Rev. Lett.*, **23**, 53–56.
Ackerman, S. A. and Stephens, G. L. (1987). The absorption of solar radiation by cloud droplets: An application of anomalous diffraction theory. *J. Atmos. Sci.*, **44**, 1574–1588.
Acquista, Ch. (1976). Light scattering by tenuous particles: A generalization of the Rayleigh–Gans–Rokard approximation. *Appl. Opt.*, **15**, 2932–2936.
Aden, A. L. and Kerker, M. (1951). Scattering of electromagnetic waves from two concentric spheres. *J. Appl. Phys.*, **22**, 1242–1246.
Adey, A. W. (1956). Scattering of electromagnetic waves by coaxial cylinders. *Can. J. Phys.*, **35**, 510–520.
Alvarez-Estrada, R. F. and Calvo, M. L. (1981). Inhomogeneous dielectric slabs with analytic frequency-dependent permittivities. *Opt. Acta*, **28**, 1253–1271.
Alvarez-Estrada, R. F., Calvo, M. L. and Juncos del Egido, P. (1980). Scattering of TM waves by dielectric fibres Iterative and eikonal solutions. *Opt. Acta*, **27**, 1367–1378.
Aragón, S. R. and Elwenspoek, M. (1982). Mie scattering from thin spherical bubbles. *J. Chem. Phys.*, **77**, 3406–3413.
Arnush, D. (1964). Electromagnetic scattering from a spherical nonuniform medium, Part I: General theory. *IEEE Trans. Antenn. Propagat.*, **AP-12**, 86–90.
Asano, S. and Yamamoto, G. (1975). Light scattering by a spheroidal particle. *Appl. Opt.*, **14**, 29–49.
Attard, P., Box, M. A., Bryant, G. and McKeller, B. H. J. (1986). Asymptotic behavior of the Mie-scattering amplitude. *J. Opt. Soc. Am.*, **A3**, 256–258.
Babenko, V. A., Astafyeva, L. G. and Kuzmin, V. N. (2003). *Electromagnetic Scattering in Disperse Media*. Springer/Praxis, Chichester, UK.

Baker, A. (1964). Relativistic high-energy approximation for elastic scattering of Dirac particles. *Phys. Rev.*, **134**, B240–B251.
Baker, A. (1972). Second order eikonal approximation for potential scattering. *Phys. Rev.*, **D6**, 3462–3469.
Baker, A. (1973). Corrections to Glauber scattering amplitude. *Phys. Rev.*, **D8**, 1937–1939.
Banerjee, H. and Mallik, S. (1974). Question of the validity of the eikonal approximation. *Phys. Rev.*, **D9**, 956–962.
Banerjee, H. and Sharma, S. K. (1978). Infrared and high energy fixed angle domains of scattering and the role of the eikonal approximation. *Ann. Phys. (N.Y.)*, **117**, 447–470.
Banerjee, H., Dutta-Roy, B. and Sharma, S. K. (1975). Corrections to the Glauber multiple scattering series. *Ann. Phys. (N.Y.)*, **95**, 127–138.
Banerjee, H., Mallik, S. and Sharma, S. K. (1977). The role of eikonal approximation in the infrared domain. *Phys. Lett.*, **B66**, 239–242.
Baran, A. J. and Havemann, S. (2000). Comparison of electromagnetic theory and various approximations for computing the absorption efficiency and single-scattering albedo of hexagonal columns. *Appl. Opt.*, **39**, 5560–5568.
Barber, P. W. and Hill, S. C. (1990). Scattering by axisymmetric particles: T-matrix method. *Light Scattering by Particles: Computational Methods* (Vol. 2 of Advanced Series in Applied Physics). World Scientific, Singapore.
Barber, P. W. and Wang, D. S. (1978). Rayleigh–Gans–Debye applicability to scattering by non-spherical particles. *Appl. Opt.*, **17**, 797–803.
Bayvel, L. P. and Jones, A. R. (1981). *Electromagnetic Scattering and Its Applications*. Applied Science, London.
Belafhal, A., Ibnchaikh, M. and Nassim, K. (2002). Scattering amplitude of absorbing and non-absorbing spheroidal particles in the WKB approximation. *J. Quant. Spectrosc. Radiat. Transf.*, **72**, 385–402.
Berlad, G. (1971). An impact-parameter representation for all scattering angles. *Nuovo Cim.*, **A6**, 594–600.
Berriman, K. B. J. and Castillejo, L. (1973). Comparison of eikonal amplitudes for potential scattering. *Phys. Rev.*, **D8**, 4647–4652.
Berry, M. and Percival, I. (1986). Optics of fractal clusters such as smoke. *Opt. Acta*, **33**, 577–591.
Bhandari, R. (1986). Tiny core or thin layer as a perturbation in scattering by a single-layered sphere. *J. Opt. Soc. Am.*, **A3**, 319–328.
Bohren, C. F. and Huffman, D. R. (1983). *Absorption and Scattering of Light by Small Particles*. John Wiley & Sons, New York.
Bohren, C. F. and Nevitt, T. J. (1983). Absorption by a sphere, a simple approximation. *Appl. Opt.*, **22, 774–775.**
Born, M. and Wolf, E. (1970). *Principles of Optics*. Pergamon Press, Oxford, UK.
Boron, S. and Waldie, B. (1978). Particle sizing by forward lobe scattered intensity–ratio technique: Errors introduced by applying diffraction theory in the Mie regime. *Appl. Opt.*, **17**, 1644–1648.
Borovoi, A. G. (1988). Approximation of straight rays in problems of wave scattering and propagation in random media. *Atmos. Opt.*, **1**, 17–21.
Borovoi, A. G. and Krutikov, V. A. (1976). Statistics of a wave field when propagating in a system of large optically soft scatterers, *Opt. Spectrosc. (USSR)*, **40**, 416–419.
Borovoi, A. G., Natts, E. I. and Oppel, U. G. (1998). Scattering of light by a red blood cell. *J. Biomed. Opt.*, **3**, 364–372.

Bourrely, C., Chiappetta, P. and Torrésani, B. (1986a). Light scattering by particles of arbitrary shape: A fractal approach. *J. Opt. Soc. Am.*, **A3**, 250–255.

Bourrely, C., Torrésani, B. and Chiappetta, P. (1986b). Scattering of an electromagnetic wave by an irregularly shaped object. *Opt. Commun.*, **58**, 365–368.

Bourrely, C., Chiappetta, P. and Marie, T. (1989). Electromagnetic scattering by large rotating particles in the eikonal formalism. *Opt. Commun.*, **70**, 173–176.

Bourrely, C., Lemaire, T. and Chiappetta, P. (1991). A vectorial description of the electromagnetic scattering by large bodies of spherical shape. *J. Mod. Opt.*, **38**, 305–315.

Bourrely, C., Chiappetta, P. and Lemaire, T. (1996). Improved version of the eikonal model for absorbing spherical particles, *J. Mod. Opt.*, **43**, 409–415.

Box, M. A. and McKeller, B. H. J. (1978). Analytic inversion of multispectral extinction data in the anomalous diffraction approximation. *Opt. Lett.*, **3**, 91–93.

Box, M. A. and McKeller, B. H. J. (1981). Further relations between analytic inversion formulas for multispectral extinction data. *Appl. Opt.*, **20**, 3829–3831.

Bricaud, A. and Morel, A. (1986). Light attenuation and scattering by phytoplanktonic cells: A theoretical modeling. *Appl. Opt.*, **25**, 571–80.

Bricaud, A., Zaneveld, J. R. V. and Kitchen, J. C. (1992). Backscattering efficiency of coccolithophorids: Use of a three layered sphere model. In: Gilbert, G. D. (ed.), *Ocean Optics XI. Proc. SPIE*, **1750**, 27–23.

Brinkmeyer, E. (1978). Refractive-index profile determination of optical fibres by spatial filtering. *Appl. Opt.*, **17**, 14–15.

Bruns, H. (1895). Das eikonal. *Abh. Kgl. Sachs. Ges. Wiss., Math.-Phys. Kl.*, **21**, 325–335.

Bryant, F. D. and Latimer, P. (1969). Optical efficiencies of large particles of arbitrary shape and orientations. *J. Coll. Interf. Sci.*, **30**, 291–304.

Burberg, R. (1956). Die Beugung electromagnetischer Wellen am unendlich langen Kreiszylinder. *Z. Naturforschung.*, **11A**, 800–806, 4823.

Byron, F. W. and Joachain, C. J. (1973). Remarkable properties of the eikonal approximation. *Physica*, **66**, 33–42.

Byron, F. W. and Joachain, C. J. (1977). Eikonal theory of electron – and positron – atom collisions. *Phys. Rep.*, **34**, 233–324.

Byron, F. W., Krotkov, R. V. and Medeiros, J. A. (1970). Quenching of metastable hydrogen. *Phys. Rev., Lett.*, **24**, 83–86.

Byron, F. W., Joachain, C. J. and Mund, E. H. (1973). Potential scattering in the eikonal approximation. *Phys. Rev.*, **D8**, 2622–2639.

Byron, F. W., Joachain, C. J. and Mund, E. H. (1975). Potential scattering in the eikonal approximation. II. *Phys. Rev.*, **D11**, 1162–1672.

Calvo, M. L. and Juncos del Egido, P. (1979). Eikonal approximation for electromagnetic wave scattering by a cladded optical fibre. *SPIE Proc.*, **213**, 35–37.

Calvo, M. L. and Juncos del Egido, P (1982). Corrections to Raman–Nath diffraction by volume holograms. *Opt. Acta*, **29**, 1061–1072.

Champion, J. V., Meeten, G. H. and Senior, M. (1979). Refraction by spherical colloidal particles. *J. Coll. Interf. Sci.*, **72**, 471–482.

Chang, T. N., Poe, R. T. and Ray, P. (1973). Glauber-theory approach for molecular vibrational excitations. *Phys. Rev. Lett.*, **31**, 1097–1099.

Chen, T. W. (1974). Approximation method for high-energy scattering at large angles. *Lett. Nuovo Cim.*, **11**, 315–320.

Chen, T. W. (1984). Generalized eikonal approximation. *Phys. Rev.*, **C30**, 585–592.

Chen, T. W. (1987). Scattering of light by a stratified sphere. *Appl. Opt.*, **26**, 4155–4158.

Chen, T. W. (1988). Eikonal approximation method for small-angle light scattering. *J. Mod. Opt.*, **35**, 743–752.

Chen, T. W. (1989). High energy light scattering in the generalized eikonal approximation. *Appl. Opt.*, **28**, 4096–4102.

Chen, T. W. (1990). Sum rules for multiple scattering of light by neighbouring dielectric spheres. *J. Appl. Phys.*, **67**, 7147–7148.

Chen, T. W. (1993). Simple formula for light scattering by a large spherical dielectric. *Appl. Opt.*, **32**, 7568–7571.

Chen, T. W. (1994). Diffraction by a spherical dielectric at large size parameter. *Opt. Commun.*, **107**, 189–192.

Chen, T. W. (1995). Effective sphere for spheroid in light scattering. *Opt. Commun.*, **114**, 199–202.

Chen, T. W. and Hoock, D. W. (1975). Backward potential scattering at high energies. *Phys. Rev.*, **D12**, 1765–1771.

Chen, T. W. and Smith, W. S. (1992). Large-angle light scattering at large size parameters. *Appl. Opt.*, **31**, 6558–6560.

Chen, T. W. and Yang, L. M. (1996). Simple formula for small-angle light scattering by a spheroid. *Opt. Commun.*, **123**, 437–442 (results reproduced with permission from Elsevier).

Chen, Z., Taflove, A. and Backman, V. (2003). Equivalent volume averaged light scattering behaviour of randomly inhomogeneous dielectric spheres in the resonant range. *Opt. Lett.*, **28**, 764–767.

Chen, Z., Taflove, A. and Backman, V. (2004). Concept of the equiphase sphere for light scattering by nonspherical dielectric particles. *J. Opt. Soc. Am.*, **21**, 88–97.

Chernyshev, A. V., Maltsev, V. P., Prots, V. I. and Doroshkin, A. A. (1995). Measurement of scattering properties of individual particles with a scanning flow cytometer. *Appl. Opt.*, **34**, 6301–6305.

Chiappetta, P. (1980). A new model for scattering by irregular absorbing particles. *Astron. Astrophys.*, **83**, 348–353.

Chýlek, P. and Klett, J. D. (1991a). Extinction cross section of non-spherical particles in the anomalous diffraction approximation. *J. Opt. Soc. Am.*, **A8**, 274–281.

Chýlek, P. and Klett, J. D. (1991b). Absorption and scattering of electromagnetic radiation by prismatic columns: Anomalous diffraction approximation. *J. Opt. Soc. Am.*, **A8**, 1713–1720.

Chýlek, P. and Li, J. (1995). Light scattering by small particles in an intermediate region. *Opt. Commun.*, **117**, 389–394.

Chýlek, P. and Videen, G. (1994). Longwave radiative properties of polydispersed hexagonal ice crystals. *Atmos. Sci.*, **51**, 175–190.

Cowley, J. M. and Moodie, J. F. (1957). The scattering of electrons by atoms and crystals. I: A new theoretical approach. *Acta Cryst.*, **10**, 609–619.

Cross, D. A. and Latimer, P. (1970). General solutions for extinction and absorption efficiencies of arbitrarily oriented cylinders by anomalous-diffraction approximation. *J. Opt. Soc. Am.*, **60**, 904–907.

Debi, S. and Sharma, S. K. (1979). Investigation of domains of validity of approximation methods in light scattering from spherical obstacles. *Opt. Acta*, **26**, 297–300.

Debye, P. (1915). Scattering from non-crystalline substances, *Ann. Physik.*, **46**, 809–823.

Deirmendjian, D. (1969). *Scattering on Spherical Polydispersions*. Elsevier, New York.

Di Marzio, F. and Szajman, J. (1992). Mie scattering in the first-order corrected eikonal approximation. *Computer Physics Commun.*, **70**, 297–304

Draine, B. T. (1988). The discrete-dipole approximation and its applications to interstellar graphite grains. *Astrophys. J.*, **333**, 848–872.
Draine, B. T. and Flatau, P. J. (1994). Discrete dipole approximation for scattering calculations. *J. Opt. Soc. Am.*, **11A**, 1491–1499.
Elderton, W. P. and Johnson, N. L. (1969). *System of Frequency Curves*. Cambridge University Press, Cambridge, UK.
Englert, F., Nicoletopoulos, P., Brout, R. and Truffin, C. (1969). *Nuovo Cim.*, **A64**, 561.
Evans, B. T. N. and Fournier, G. R. (1990). A simple approximation to extinction efficiency valid over all size parameters. *Appl. Opt.*, **29**, 4666–4670.
Evans, B. T. N. and Fournier, G. R. (1994). Analytic approximation to randomly oriented spheroid extinction. *Appl. Opt.*, **33**, 5796–5804.
Farafonov, V. G. (1983). Difraktsiya ploskoj e'lektromagnitnoj volny na die'lektricheskom sferoide. *Differents. Uravn.*, **19**, 1765–1777.
Farafonov, V. G., Il'in, V. B. and Prokopjeva, M. S. (2001). Scattering of light by homogeneous and multilayered ellipsoids in quasistatic approximation. *Opt. Spectrosc.*, **92**, 608–617.
Farone, W. A. (1965). Generalization of Rayleigh–Gans scattering from radially inhomogeneous spheres. *J. Opt. Soc. Am.*, **55**, 737–738.
Farone, W. A. and Robinson, M. J. I. (1968). The range of validity of the anomalous diffraction approximation to electromagnetic scattering by a sphere. *Appl. Opt.*, **7**, 643–645.
Farone, W. A., Kerker, M. and Matijevic, E. (1963). Scattering by infinite cylinders at perpendicular incidence. In: M. Kerker (ed.), *Electromagnetic Scattering* (pp. 55–71). Pergamon Press, London.
Fiel, R. J. (1970). Small angle light scattering of bioparticles, I: Model systems. *Exp. Cell Res.*, **59**, 413.
Flatau, P. J. (1992). *Scattering by Irregular Particles* (Paper No. 517, Ph.D thesis). Department of Atmospheric Science, Colorado State University, CO.
Flugge, S. (1971). *Practical Quantum Mechanics* (Vol. 1). Springer-Verlag, Berlin.
Fournier, G. R. and Evans, B. T. N. (1991). Approximation of extinction efficiency for randomly oriented spheroids. *Appl. Opt.*, **30**, 2042–2048.
Fournier, G. R. and Evans, B. T. N. (1996). Approximations to extinction from randomly oriented circular and elliptic cylinders. *Appl. Opt.*, **35**, 4271–4282.
Franco, V. (1968). Diffraction theory of scattering by hydrogen atoms. *Phys. Rev. Lett.*, **20**, 709–712.
Franssens, F., De Maziere, M. and Fonteyn, D. (2000). Determination of aerosol size distribution by analytic inversion of the extinction spectrum in the complex anomalous diffraction approximation. *Appl. Opt.*, **39**, 4214–4231.
Fu, Q., Sun, W. B. and Yang, P. (1999). Modelling of scattering and absorption by nonspherical cirrus ice particles at thermal infrared wavelengths. *J. Atmos. Sci.*, **56**, 2937–2947.
Fymat, A. L. (1978). Analytical inversions in remote sensing of particle size distributions. 1: Multispectral extinctions in the anomalous diffraction approximation. *Appl. Opt.*, **17**, 1675–1676.
Fymat, A. L. and Mease, K. D. (1981). Mie forward scattering: Improved semiempirical approximation with application to particle size distribution inversion. *Appl. Opt.*, **20**, 194–198.
Fymat, A. L. and Smith, C. B. (1979). Analytical inversion of remote sensing of particle size distributions. 4: Comparison of Fymat and Box–McKeller solutions in the anomalous diffraction approximation. *Appl. Opt.*, **18**, 3595–3598.
Gans, R. (1925). Strahlungsdiagramme ultramikroskopisher Teilchen. *Ann. Phys.*, **76**, 29–38.

References

Garibaldi, U., Levi, A. C., Spadacini, R. and Tommei, G. E. (1975). Quantum theory of atom–surface scattering: Diffraction and rainbow. *Surface Sci.*, **48**, 649–675.

Gerjuoy, E. and Thomas, B. K. (1974). Applications of Glauber approximation to atomic collisions. *Rep. Prog. Phys.*, **37**, 1345–1431.

Gersten, J. I. and Mittileman, M. H. (1975) Eikonal theory of charged particle scattering in the presence of a strong electromagnetic wave. *Phys. Rev.*, **A12**, 1840–1845.

Gien, T. T. (1988) The modified Glauber approximation. *Phys. Reports*, **160**, 123–187.

Giese, R. H., Weiss, K., Zerull, R. H. and Ono, T. (1978). Large fluffy particles: A possible explanation of the optical properties of interplanetary dust. *Astron. Astrophys.*, **65**, 265–272.

Glauber, R. J. (1959). High energy collision theory. In: W. E. Brittin and L. G. Dunham (eds), *Theoretical Physics* (Vol. I, pp. 315–414). Interscience, New York.

Gómez, A. and Castaño, V. M. (1988). Unified approach to the high-energy approximation in transmission electron microscopy. *Phys. Status. Solidi*, **A107**, 845–850.

Gordon, J. E. (1985). Simple method for approximating Mie scattering. *J. Opt. Soc. Am.*, **A2**, 156–159.

Gradshteyn, I. S. and Ryzhik, I. M. (1980). *Tables of Integrals, Series and Products*. Academic Press, New York.

Granovskii, Ya. I. and Stón, M. (1994a). Light scattering cross–sections: Summing of Mie–series. *Physica Scripta*, **50**, 140–141.

Granovskii, Ya. I. and Stón, M. (1994b). Attenuation of light scattered by transparent particles. *JETP(USA)*, **78**, 645–649.

Greenberg, J. M. (1960). Scattering by nonspherical particles. *J. Appl. Phys.*, **31**, 82–84.

Güttler, A. (1952). Die Mie theorie der Beugung durch dielecktrische Kugeln mit absorbierendem kern und ihre Bedeutung für Probleme der interstellaren Materie und des atmospharischen Aerosols. *Ann. der Phys.*, **11**, 65–98.

Hahn, Y. (1970). Impact parameter amplitudes for large-angle potential scattering, II. *Phys. Rev.*, **C2**, 775–781.

Haltrin, V. I. (2002). One-parameter two-term Henyey–Greenstein phase function for light scattering in seawater. *Appl. Opt.*, **41**, 1022–1028.

Hamilton, W. R. (1828). Theory of system of rays. *Trans. R. Ir. Acad.*, **15**, 69–84.

Hammer, M., Schweitzer, D., Michel, B., Thamm, E. and Kolb, A. (1998). Single scattering by red blood cells. *Appl. Opt.*, **37**, 7410–7418.

Harnad, J. P. (1975). The eikonal approximation and the E(2) invariance. *Ann. Phys. (NY)*, **91**, 413–414.

Hart, R. W. and Montroll, E. W. (1951). On the scattering of plane waves by soft particles, I: Spherical obstacles. *J. Appl. Phys.*, **22**, 376–386.

He, J., Karlsson, A., Swartling, J. and Andersson-Engels, S. (2004). Light scattering by multiple red blood cells. *J. Opt. Soc. Am.*, **A21**, 1953–1961.

Heller, W. (1963). Theoretical and experimental investigation of light scattering colloidal spheres. In: M. Kerker (ed.), *Electromagnetic Scattering* (pp. 107–120). Pergamon Press, London.

Heller, W., Nakagaki, M. and Wallach, M. L. (1959). Theoretical investigations on the light scattering of colloidal spheres, V: Forward scattering. *J. Chem. Phys.*, **30**, 444–450.

Henyey, L. and Greenstein, J. (1941). Diffuse radiation in the galaxy. *Astrophys. J.*, **93**, 70–83.

Hodgkinson, R. J. (1966). Particle sizing by means of forward scattering lobe. *Appl. Opt.*, **5**, 839–844.

Holt, A. R. and Shepherd, J. W. (1979). Electromagnetic scattering by dielectric spheroids in the forward and backward directions. *J. Phys. (UK)*, **A12**, 159–166.

Huang, Z., Chidichimo, G., Nicoletta, F. P., De Simone, B. C. and Caruso, C. (1996). A model of an aligned nematic droplet for small angle light scattering. *J. Appl. Phys.*, **80**, 6155–6159.

Hunziker, W. (1963). Potential scattering at high energies. *Helv. Phys. Acta*, **36**, 838–856.

Irvine, W. M. (1963). The asymmetry parameter of the scattering diagram of a spherical particle. *Bull. Astron. Inst. Netherlands*, **3**, 176–184.

Ishimaru, A. (1997). *Wave Propagation and Scattering in Random Media*. IEEE Press, New York.

Joachain, C. J. (1975). *Quantum Collision Theory* (chaps 8 and 9). North Holland, Amsterdam.

Joachain, C. J. and Quigg, C. (1974). Multiple scattering expansions in several particle dynamics. *Rev. Mod. Phys.*, **46**, 279–324.

Jobst, G. (1925). *Ann. Phys. Lpz.*, **78**, 157.

Jones, A. R. (1999). Light scattering for particle characterization. *Progr. Energ. Combust. Sci.*, **25**, 1–53.

Jones, A. R., Koh, J. and Nassaruddin, A. (1996). Error contour charts for the two-wave WKB approximation. *J. Phys. D (UK)*, **29**, 39–42.

Jones, D. S. (1957). High-frequency scattering of electromagnetic waves. *Proc. R. Soc. London Ser. A*, **240**, 206–213.

Jung, Y. D. (1996). Eikonal differential scattering cross sections for elastic electron–ion collisions in strongly coupled plasmas. *Phys. Plasmas*, **3**, 4376–4379.

Kahnert, F. M. (2003). Numerical methods in electromagnetic scattering theory. *J. Quant. Spectrosc. Radiat. Transf.*, **79–80**, 775–824.

Katz, A., Alimova, A., Xu, M., Gottlieb, P., Rudolph, E., Shah, M. K., Savage, H. E., Rosen, R. B., McCormick, S. A. and Alfano, R. R. (2003). Bacteria size determination by elastic light scattering. *IEEE J. Sel. Top. Quantum Electron.*, **9**, 277–287.

Katz, A., Alimova, A., Xu, M., Gottlieb, P., Rudolph, E., Steiner, J. C. and Alfano, R. R. (2005). *In situ* determination of refractive index and size of *Bacillus* spores by light transmission. *Opt. Lett.*, **30**, 589–591.

Kerker, M. (1969). *The \sim cattering of Light and Wther Electromagnetic Radiation*. Academic Press, New York.

Kerker, M. and Matijević, E. (1961). Scattering of electromagnetic waves from concentric infinite cylinders. *J. Opt. Soc. Am.*, **51**, 506–509.

Kerker, M., Farone, W. A. and Matijević, E. (1963). Applicability of Rayleigh–Gans scattering to spherical particles. *J. Opt. Soc. Am.*, **53**, 758–759.

Khlebtsov, N. G. (1993). Optics of fractal clusters in the anomalous diffraction approximation. *J. Mod. Opt.*, **40**, 2221–2235.

Kim, C., Lior, N. and Okuyama, K., (1996). Simple mathematical expressions for spectral extinction and scattering properties of small size-parameter particles, including examples for soot and TiO_2. *J. Quant. Spectrosc.Rad. Transf.*, **55** 391–411.

Kitchen, J. C. and Zaneveld, J. R. V. (1992). A three-layer sphere model of the optical properties of phytoplankton. *Limnol. Oceanogr.*, **37**, 1680–1690.

Klett, J. D. (1984). Anomalous diffraction model for inversion of multispectral extinction data including absorption effects. *Appl. Opt.*, **23**, 4499–4508.

Klett, J. D. and Sutherland, R. A. (1992). Approximate methods for modeling the scattering properties of non-spherical particles: Evaluation of the Wentzel–Kramers–Brillouin method. *Appl. Opt.*, **31**, 373–386.

Koch, A. L. (1968). Theory of the angular dependence of light scattered by bacteria and similar sized biological objects. *J. Theor. Biol.*, **18**, 133.

Kocifaj, M. (2004). Interstellar dust extinction problem: Benchmark of (semi) analytic approaches and regularization method. *Contrib. Astron. Obs. Skalnaté Pleso.*, **34**, 141–156.

Kokhanovsky, A. A. (1995). About edge effects in light absorption by weak absorbing particles. *Opt. Spectrosc.*, **78**, 967–969.

Kokhanovsky, A. A. (2005). *Light Scattering Media Optics: Problems and Solutions*. Springer-Verlag, London.

Kokhanovsky, A. A. and Zege, E. P. (1995). Local optical parameters of spherical polydispersions: Simple approximations. *Appl. Opt.*, **34**, 5513–5519.

Kokhanovsky, A. A. and Zege, E. P. (1997). Optical properties of aerosol particles: A review of approximate analytic solutions. *Aerosol Sci.*, **28**, 1–21 (Figure 4.6 reproduced with permission from Elsevier).

Kujawaski, E. (1971). Validity of eikonal-type approximations for potential scattering. *Phys. Rev.*, **D4**, 2573–2577.

Kujawaski, E. (1972). Multiple scattering with a linearized propagator. *Ann. Phys. (NY)*, **74**, 567–594.

Kuznetsov, V. V. and Pavlova, L. N. (1988). Attenuation and absorption of radiation by optically soft cylindrical particles. *Bull. USSR Acad. Sci. Atmos. Oceanic Phys.*, **24**, 147–151.

Latimer, P. (1975). Light scattering by ellipsoids, *J. Coll. Interf. Sci.*, **53**, 102–109.

Latimer, P. (1980) Predicted scattering by spheroids: Comparison of approximate and exact methods. *Appl. Opt.*, **19**, 3039–3041.

Latimer, P. (1984a). Light scattering by a homogeneous sphere with radial projections. *Appl. Opt.*, **23**, 442–447.

Latimer, P. (1984b). Light scattering by a structured particle: The homogeneous sphere with holes. *Appl. Opt.*, **23**, 1844–1847.

León, J., Quirós, M. and Mittelbrunn, J. R. (1977). A group theoretical approach to relativistic eikonal physics. *Nuovo Cim.*, **41A**, 141–165.

Lévy, M. and Sucher, J. (1969). Eikonal approximation in quantum field theory. *Phys. Rev.*, **186**, 1656–1670.

Lin, F. C. and Fiddy, M. A. (1992). Born–Rytov controversy, I: Comparing analytical and approximate expressions for the one dimensional deterministic case. *J. Opt. Soc. Am.*, **A9**, 1102–1110.

Lind, A. C. and Greenberg, J. M. (1966). Electromagnetic scattering by obliquely oriented spheroids. *J. Appl. Phys.*, **37**, 3195–3203.

Liou, K. N. and Takano, Y. (1994). Light scattering by non-spherical particles: Remote sensing and climatic implications. *Atmos. Res.*, **31**, 271–298.

Liu, C. (1998). Validity of anomalous diffraction approximation in $m - \chi$ domain. *Atmos. Res.*, **49**, 81–86.

Liu, C., Jonas, P. R. and Saunders, C. P. R. (1996). Accuracy of the anomalous diffraction approximation to light scattering by column-like ice crystals. *Atmos. Res.*, **41**, 63–69.

Liu, Y., Arnott, W. P. and Hallet, J. (1998). Anomalous diffraction theory for arbitrarily oriented finite circular cylinders and comparison with exact T-matrix results. *Appl. Opt.*, **37**, 5019–5029.

Logan, N. A. (1965). Survey of some early studies of the scattering of plane waves by a sphere. *Proc. IEEE*, 773–785.

Lopatin, V. N. and Sid'ko, Ya. F. (1987). Absorption of electromagnetic radiation by "soft" structured particles. *Izv. Atmos. Oceanic Phys.*, **23**, 396–401.

Lopatin, V. N. and Sidko, Ya. F. (1988). *Introduction to Optics of Cell Suspension*. Nauka, Moscow.

Lui, C. W., Clarkson, M. and Nicholls, R. W. (1996). An approximation for spectral extinction of atmospheric aerosols. *J. Quant. Spectrosc. Radiat. Transf.*, **55**, 519–531.

Mahood, R. W. (1987). The application of vector diffraction to the scalar anomalous diffraction approximation of van de Hulst. Masters thesis, Pennsylvania State University, Department of Meteorology.

Maltsev, V. P., Chernyshev, A. V., Sem'yanov, K. A. and Soini, E. (1996). Absolute real-time measurement of particle size distribution with the flying light-scattering indicatrix method. *Appl. Opt.*, **35**, 3275–3280.

Mano, Y. (2000). Exact solution of electromagnetic scattering by a three-dimensional hexagonal ice column obtained with boundary element method. *Appl. Opt.*, **39**, 5541–5546.

Marburger, J. H. and Felber, F. S. (1978). Theory of a lossless non-linear Fabry–Perot interferometer. *Phys. Rev.*, **A17**, 335–342.

Maslowska, A. (1991). Interaction of light with particles. *Acta Geophys. Polonica*, **39**, 113–128.

Maslowska, A., Flatau, P. J. and Stephens, G. L. (1994). On the validity of anomalous diffraction theory to light scattering by cubes. *Opt. Commun.*, **107**, 35–40.

McKeller, B. H. J. (1982). Light scattering determination of size distribution of cylinders: An analytic approximation. *J. Opt. Soc. Am.*, **72**, 671–672.

Meeten, G. H. (1979) Induced birefringence in colloidal dispersions. *J. Chem. Phys. (Faraday-Trans. II)*, **75**, 1406–1415.

Meeten, G. H. (1980a). The intrinsic optical anisotropy of colloidal particles in the anomalous diffraction approximation. *J. Coll. Interf. Sci.*, **74**, 181–185.

Meeten, G. H. (1980b). Refractive index of colloidal dispersions of spheroidal particles. *J. Coll. Interf. Sci.*, **77**, 1–5.

Mie, G. (1908). Beiträge zur Optik truber Medien speziell kolloidaler Metallösungen. *Ann. Phys.*, **25**, 377–445.

Mikulski, J. J. and Murphy, E. L. (1963). The computation of electromagnetic scattering from concentric spherical structures. *IEEE Trans. Antenn. Propagat.*, **AP-11**, 169–177.

Min, M., Hovenier, J. W., Diminik, C., de Koter, A. and Yurkin, M. A. (2006). Absorption and scattering properties of arbitrarily shaped particles in the Rayleigh domain: A rapid computational method and a theoretical foundation for statistical approach. *J. Quant. Spectrosc. Rad. Transf.*, **97**, 161–180 (theorem reprinted with permission from Elsevier).

Mishchenko, M. I., Wiscombe, W. J., Hovenier, J. W. and Travis, L. D. (2000). Overview of scattering by non-spherical particles. In: M. I. Mishchenko, J. W. Hovenier and L. D. Travis (eds), *Light Scattering by Nonspherical Particles* (pp. 29–60). Academic Press, San Diego, CA.

Mishchenko, M. I., Travis, L. D. and Lacis, A. A. (2002). *Scattering, Absorption, and Emission of Light by Small Particles.* Cambridge University Press, Cambridge, UK.

Moeglich, F. (1927). Beugungserscheinungen an Korpen von ellipsoidischer Gestalt. *Ann. Phys.*, **83**, 609–735.

Molière, G. (1947). Single scattering in a screened Coulomb field: Theory of scattering of fast charged particles. *Z. Naturforsch.*, **A2**, 133–145.

Montroll, E. W. and Greenberg, J. M. (1952). Scattering of plane waves by soft obstacles, III: Scattering by obstacles with spherical and circular symmetry. *Phys. Rev.*, **86**, 889–897.

Montroll, E. W. and Hart, R. W. (1951). On the scattering of plane waves by soft particles, II: Scattering by cylinders, spheroids and disks. *J. Appl. Phys.*, **22**, 1278–1289.

Moore, R. J. (1970). High energy approximation in potential scattering theory. *Phys. Rev.*, **D2**, 313–316.

Morris, V. J. and Jennings, B. R. (1977). Anomalous diffraction approximation to the low-angle light scattering from coated spheres. *Biophys. J.*, **17**, 95–101.

Newton, R. G. (1966). *Scattering Theory of Waves and Particles.* McGraw-Hill, New York.

Nicholls, R. W. (1984). Wavelength dependent spectral extinction of atmospheric aerosols. *Appl. Opt.*, **23**, 1142–1143.

Nussenzveig, H. M. (1984). M. Schönberg on his 70th birthday. *Rev. Bras. Fis.*, Special Volume, 302–319.

Nussenzveig, H. M. and Wiscombe, W. J. (1980). Efficiency factors in Mie scattering. *Phys. Rev. Lett.*, **45**, 1490–1494.

Orenstein, M., Speiser, S. and Katriel, J. (1984). An eikonal approximation for non-linear resonators exhibiting bistability. *Opt. Commun.*, **48**, 367–373.

Orenstein, M., Speiser, S. and Katriel, J. (1985). A general eikonal treatment of coupled dispersively non-linear resonators exhibiting optical multistability. *IEEE J. Quantum Electron.*, **QE-21**, 1513–1522.

Orenstein, M., Katriel, J. and Speiser, S. (1986). In: Y. Prior (ed.), *Method of Laser Spectroscopy*. Plenum Press, New York.

Orenstein, M., Katriel, J. and Speiser, S. (1987a). Optical bistability in molecular systems exhibiting non-linear absorption. *Phys. Rev.*, **A35**, 1192–1209.

Orenstein, M., Katriel, J. and Speiser, S. (1987b). Optical bistability in molecular systems exhibiting non-linear absorption. *Phys. Rev.*, **A35**, 2175–2183.

Paramonov, L. E. (1994) On optical equivalence of randomly oriented ellipsoidal and polydisperse spherical particles: The extinction, scattering and absorption cross-sections. *Opt. Spectrosc.*, **77**, 660–663.

Paramonov, L. E. (1995). Evaluation of absorption cross-section of polydisperse chaotically oriented soft ellipsoidal particles. *Optika Spektrokopiya*, **78**, 964–966.

Paramonov, L. E. (1996). Theoretical analysis of optical spectra of algal absorption. *Oceanology*, **35**, 655–659 [English translation].

Paramonov, L. E., Lopatin, V. N. and Sidko, F. Ya. (1986). Light scattering by soft spheroidal particles. *Opt. Spectrosc.*, **61**, 570–576.

Penndorf, R. (1960). *Scattering Coefficients for Absorbing and Non-absorbing Aerosols* (Technical Report RAD-TR-60-27). Air Force Cambridge Res. Lab., Bedford, MA.

Penndorf, R. (1962). Scattering and absorption coefficients for small absorbing and non-absorbing aerosols. *J. Opt. Soc. Am.*, **52**, 896–904.

Perelman, A. Y. (1978). An application of Mie series to soft particles. *Pure Appl. Geophys.* (Birkhäuser Publishing, Basel, Switzerland), **116**, 1077–1088.

Perelman, A. Y. (1985). The scattering of light by a translucent sphere described in soft-particle approximation, *Dokl. Akad. Nauk SSSR.*, **281**, 51–54.

Perelman, A. Y. (1991) Extinction and scattering by soft particles. *Appl. Opt.*, **30**, 475–484.

Perelman, A. Y. (1994). Improvement of convergence of a series for soft sphere absorption cross section. *Opt. Spectrosc.*, **77**, 643–647.

Perelman, A. Y. (1997). Integral representation of fields and their spectra in the Mie problems (S-approximation). *Opt. Spectrosc.*, **82**, 423–433.

Perelman, A. Y. and Voshchinnikov, N. V. (2002). Improved S-approximation for dielectric particles. *J. Quant Spectrosc Rad. Transf.*, **72**, 607–621.

Perelman, L. T., Backman, V., Wallace, M., Zonios, G., Manoharan, R., Nusrat, A., Shields, S., Seiler, M., Lima, C., Hamano, T. *et al.* (1998). Observation of periodic fine structure in reflectance from biological tissue: A new technique for measuring nuclear size distribution. *Phys. Rev. Lett.*, **80**, 627–630.

Perrin, J. M. and Chiappetta, P. (1985). Light scattering by large particles, I: A new theoretical description of the eikonal picture. *Opt. Acta*, **32**, 907–921.

Perrin, J. M. and Lamy, P. (1983). Light scattering by large particles. *Opt. Acta*, **30**, 1223–1244.

Perrin, J. M. and Lamy, P. (1986). Light scattering by large particles, II: A vectorial description in the eikonal picture. *Opt. Acta*, **33**, 1001–1022.

Peterlin, A. and Stuart, H. A. (1939). The theory of streaming birefringence of colloids and large molecules in solution. *Z. Phys.*, **112**, 1–19.

Platzmann, P. M. and Ozaki, H. M. (1960). Scattering of electromagnetic waves from an infinitely long magnetized cylindrical plasma. *Appl. Phys.*, **31**, 1598–1601.

Posselt, B., Farafonov, V. G., Il'in, V. B., and Prokopjeva, M. S. (2002). Light scattering by multilayered ellipsoidal particles in quasistatic approximation. *Meas. Sci. Technol.*, **13**, 256–262.

Powers, S. R. and Somerford, D. J. (1979). Error contour charts relevant to fibre sizing using light scattering. *J. Phys. (UK)*, **D12**, 1809–1818.

Powers, S. R. and Somerford, D. J. (1982). Correction factors for sizing transparent fibres using light scattering. *J. Phys. (UK)*, **D15**, 403–409.

Punina, V. A. and Perelman, A. Y. (1969). Über die Berechnung der Grössenverteilung von den absorbierenden Kugelförmigen. *Pageoph*, **74**, 92–104.

Purcell, E. M. and Pennypacker, C. R. (1973). Scattering and absorption of light by non-spherical dielectric grains. *Astrophys. J.*, **186**, 705–714.

Quinby-Hunt, M. S., Hunt, A. J., Lofftus, K. and Shapiro, D. (1989). Polarized-light scattering studies of marine *Chlorella. Limnol. Oceanogr.* **34**, 1587–1600.

Quirantes, A. and Bernard, S. (2004). Light scattering by marine algae: Two-layer spherical and nonspherical models. *J. Quant. Spectrosc. Rad. Transf.*, **89**, 311–321.

Raman, C. V. and Nath, N. S. N. (1935). The diffraction of light by high frequency sound waves: Part I. *Proc. Indian Acad. Sci.*, **A2**, 406–412.

Raman, C. V. and Nath, N. S. N. (1936). The diffraction of light by high frequency sound waves: Part IV. *Proc. Indian Acad. Sci.*, **A3**, 119–125.

Rayleigh, D. W. (1881). On the electromagnetic theory of light. *Phil. Mag.*, Ser. 5, **12**, 81–101.

Rayleigh, D. W. (1914). On the diffraction of light by spheres of small refractive index. *Proc. R. Soc. London*, **A90**, 219–225.

Rayleigh, D. W. (1918). On the scattering of light by spherical shells, and by complete spheres of periodic structure, when the refractivity is small. *Proc. R. Soc. London*, **A94**, 296–300.

Reading, J. F. and Bassichis, W. H. (1972). High energy scattering at backward angles. *Phys. Rev.*, **D5**, 2031–2041.

Rosen, D. L. and Pendleton, J. D. (1995). Detection of biological particles by use of circular dichroism measurements improved by scattering theory. *Appl. Opt.*, **34**, 5875–5884.

Roy, A. K. and Sharma, S. K. (1996). On the validity of soft particle approximations for the light scattering by a homogeneous dielectric sphere. *J. Mod. Opt.*, **43**, 2225–2237. [Available at *http://www.tandf.co.uk*]

Roy, A. K. and Sharma, S. K. (1997). A new approach to inverse scattering problem. *Appl. Opt.*, **36**, 9487–9495.

Roy, A. and Sharma, S. K. (2005). A simple analysis of the extinction spectrum of a size distribution of Mie particles. *J. Opt. A: Pure Appl. Opt.* (IOP Publishing), **7**, 675–684.

Rysakov, V. M. (2004). Light scattering by "soft" particles of arbitrary shape and size. *J. Quant. Spectrosc. Rad. Transf.*, **87**, 261–287.

Rysakov, V. M. (2006). Light scattering by "soft" particles of arbitrary shape and size, II: Arbitrary orientation of particles in the space. *J. Quant. Spectrosc. Rad. Transf.*, **98**, 85–100 (results reproduced with permission from Elsevier).

Rytov, S. M. (1937). Diffraction of light by ultrasonic waves. *Izv. Akad. Nauk SSSR Ser. Fiz.*, **2**, 223.

Sarkar, S. (1980). Higher order terms in the eikonal expansion of the T-matrix for potential scattering. *Phys. Rev.*, **D12**, 3437–3458.

Saxon, D. S. (1955). *Lectures on Scattering of Light* (Scientific Report No. 9). UCLA, Los Angeles.

Saxon, D. S. and Schiff, L. I. (1957). Theory of high energy potential scattering. *Nuovo Cim.*, **6**, 614–627.

Schiff, L. I. (1956). Approximation methods for high energy potential scattering. *Phys. Rev.*, **103**, 443–453.

Schiff, L. I. (1968). *Quantum Mechanics*. McGraw Hill, New York.

Schulp, W. A. (1989). On the inversion of atomic beam scattering data in the eikonal approximation. *Surface Science*, **211/212**, 180–186.

Sharma, S. K. (1986). Density profile determination of cylindrically symmetric non-uniform plasma by spatial filtering. *Plasma Phys. Contr. Fusion* (IOP Publishing), **28**, 391–392.

Sharma, S. K. (1989). Approximate formulae for the scattering of light by oriented infinitely long homogeneous soft circular cylinders. *J. Mod. Opt.*, **36**, 399–404.

Sharma, S. K. (1992). On the validity of the anomalous diffraction approximation. *J. Mod. Opt.*, **39**, 2355–2361. [Available at *http://www.tandf.co.uk*]

Sharma, S. K. (1993). A modified anomalous diffraction approximation for intermediate size soft particles. *Opt. Commun.*, **100**, 13–18 (results used with permission from Elsevier).

Sharma, S. K. (1994). On the validity of soft particle approximations for the scattering of light by infinitely long homogeneous cylinders. *J. Mod. Opt.*, **41**, 827–838. [Available at *http://www.tandf.co.uk*]

Sharma, S. K. and Dasgupta, B. (1987). Scattering of electromagnetic waves by a non-uniform cylindrical plasma in the eikonal approximation. *Plasma Phys. Contr. Fusion* (IOP Publishing), **29**, 303–311.

Sharma, S. K. and Debi, S. (1978). Eikonal approximation to low-angle light scattering. *Biophys. J.*, **21**, 287–288.

Sharma, S. K. and Debi, S. (1980). On the accuracy of some approximation methods for weight and size determination of spherical particles by forward light scattering. *Molec. Phys.*, **40**, 1527–1530.

Sharma, S. K. and Saha, R. K. (2004). On the validity of some new acoustic scattering approximations. *Waves in Random Media*, **14**, 525–537.

Sharma, S. K. and Somerford, D. J. (1982). Approximation methods for sizing transparent fibres using light scattering. *J. Phys. (UK)* (IOP Publishing), **D15**, 2149–2156.

Sharma, S. K. and Somerford, D. J. (1983a). The eikonal approximation applied to sizing transparent fibres using the forward scattered intensity ratio technique. *J. Phys. (UK)* (IOP Publishing), **D16**, 733–742.

Sharma, S. K. and Somerford, D. J. (1983b). Investigation of domains of validity of approximation methods in forward light scattering from absorbing cylinders. *Opt. Commun.*, **45**, 1–4. [Results used with permission from Elsevier.]

Sharma, S. K. and Somerford, D. J. (1988). Modified Rayleigh–Gans–Debye approximation applied to sizing transparent homogeneous long fibres of intermediate size. *J. Phys. (UK)* (IOP Publishing), **D21**, 403–406.

Sharma, S. K. and Somerford, D. J. (1989). A comparison of extinction efficiencies in the eikonal and anomalous diffraction approximations. *J. Mod. Opt.*, **36**, 1411–1413.

Sharma, S. K. and Somerford, D. J. (1990). The eikonal approximation revisited. *Nuovo Cim.*, **D12**, 719–748.

Sharma, S. K. and Somerford, D. J. (1991). Approximation methods for characterization of particles in cohesive sediments by light scattering. *J. Phys. (UK)*, **D24**, 21–25.

Sharma, S. K. and Somerford, D. J. (1994). An approximation method for backward scattering of light by a soft spherical obstacle. *J. Mod. Opt.*, **41**, 1433–1444. [Available at *http://www.tandf.co.uk*]

Sharma, S. K. and Somerford, D. J. (1996). On the relationship between the S-approximation and Hart–Montroll approximation. *J. Opt. Soc. Am.*, **13A**, 1285–1286.

Sharma, S. K. and Somerford, D. J. (1999). Scattering of light in the eikonal approximation. *Progress in Optics*, **39**, 211–290 (results reproduced with permission from Elsevier).

Sharma, S. K., Powers, S. R. and Somerford, D. J. (1981). Investigation of domains of validity of approximation methods in light scattering from long cylinders. *Opt. Acta*, **28**, 1439–1446. [Available at *http://www.tandf.co.uk*]

Sharma, S. K., Somerford, D. J. and Sharma, S. (1982). Investigation of validity domains of corrections to eikonal approximation in forward light scattering from homogeneous spheres. *Opt. Acta*, **29**, 1677–1682. [Available at *http://www.tandf.co.uk*]

Sharma, S. K., Sharma, S. and Somerford, D. J. (1984a). The eikonal approximation applied to sizing transparent homogeneous spheres. *J. Phys. (UK)* (IOP Publishing), **D17**, 2191–2197.

Sharma, S., Sharma, S. K. and Somerford, D. J. (1984b). The eikonal approximation to near forward light scattering from homogeneous spheres. *Opt. Acta*, **31**, 867–871.

Sharma, S. K., Roy, T. K. and Somerford, D. J. (1988a). The eikonal approximation vs. the high energy approximation in optical scattering. *J. Phys. (UK)* (IOP Publishing), **D21**, 1685–1691.

Sharma, S. K., Ghosh, G. and Roy, T. K. (1988b). Effect of nature of index profile on the validity of the eikonal approximation, *J. Mod. Opt.*, **35**, 703–710. [Available at *http://www.tandf.co.uk*]

Sharma, S. K., Roy, T. K. and Somerford, D. J. (1988c). Approximate formulae for scattering of radiation from infinitely long homogeneous right circular cylinders. *J. Mod. Opt.*, **35**, 1213–1224. [Available at *http://www.tandf.co.uk*]

Sharma, S. K., Somerford, D. J. and Roy, A. K. (1997a). Simple formulae within the framework of anomalous diffraction approximation for light scattered by an infinitely long cylinder. *Pure Appl. Opt.* (IOP Publishing), **6A**, 565–575.

Sharma, S. K., Ghosh, G. and Somerford, D. J. (1997b). The S-approximation for light scattering by an infinitely long cylinder. *Appl. Opt.*, **36**, 6109–6114.

Shatilov, A. V. (1960). On the scattering of light by dielectric ellipsoids comparable to the light wavelength, I. *Opt. Spectrosc.*, **9(1)**, 86–91 [in Russian].

Shepelevich, N. V., Prostakova, I. V. and Lopatin, V. N. (1999). Extrema in the light scattering indicatrix of a homogeneous spheroid. *J. Quant. Spectrosc. Rad. Transf.*, **63**, 353–367.

Shepelevich, N. V., Prostakova, I. V. and Lopatin, V. N. (2001). Light-scattering by optically soft randomly oriented spheroids. *J. Quant. Spectrosc. Rad. Transf.*, **70**, 375–381.

Shifrin, K. S. (1952). Light scattering by two-layered particles. *Izv. AN USSR Ser. Geophys.*, **N2**, 15–21.

Shifrin, K. S. (1955). On calculations of radiative properties of water clouds. *Trudy Glavnoi Geophys. Observ.*, **46**, 5–33.

Shifrin, K. S. (1988). *Physical Optics of Ocean Water*. American Institute of Physics, New York.

Shifrin, K. S. and Perelman, A. Y. (1967). Inversion of light scattering data for the determination of spherical particle spectrum. In: R. L. Powell and R. S. Stein (eds), *Electromagnetic Scattering* (Vol. II, pp. 131–167). Gordon & Breach, New York.

Shifrin, K. S. and Punina, V. A. (1968). Light-scattering indicatrix in the region of small angles. *Bull. Izv. Acad. Sci. USSR Atmos. Oceanic Phys.*, **4**, 450–453.

Shifrin, K. S. and Stón, M. (1976). About using the RGD approximation to calculate light extinction in the ocean and atmosphere optics problems. *Izv. AN SSSR Fiz. Atm. Okeana*, **28**, 107–109.

Shifrin, K. S. and Tonna, G. (1992). Simple formula for absorption coefficient of weakly refracting particles. *Opt. Spectrosc.*, **72**, 487–490.

Shifrin, K. S. and Tonna, G. S. (1993). Inverse problems related to light scattering in the atmosphere and ocean. *Adv. Geophys.*, **34**, 175–252.

Shifrin, K. S. and Zolotov, I. S. (1993). Remark about the notation used for calculating the electromagnetic field scattered by a spherical particle. *Appl. Opt.*, **32**, 5397–5398.

Shimizu, K. (1983). Modification of the Rayleigh–Debye scattering. *J. Opt. Soc. Am.*, **73**, 504–507.

Shvalov, A. A., Soini, J. T., Chernyshev, A. V., Tarasov, P. A., Soini, E. and Maltsev, V. P. (1999). Light-scattering properties of individual erythrocytes. *Appl. Opt.*, **38**, 230–235.

Sitenko, A. G. (1959). On the theory of nuclear reactions involving complex particles. *Ukr. Fiz. Zh.*, **4**, 152–163.

Smith, C. B. (1982). Inversion of the anomalous diffraction approximation for variable complex refractive index. *Appl. Opt.*, **21**, 3363–3366.

Sommerfeld, A. and Runge, I. (1911). Anwendung der Vektorrechnung auf die Grundlagen der geometrischen Optik. *Annalen Phys.*, **35**, 277–298.

Stepanov, A. V. and Shelagin, A. V. (1986). Analysis of large scale heterogenities of a substance based on the integral cross section of elastic scatter of very cold neutrons. *Sov. Phys.-Lebedev Inst. Rep.*, **3**, 27–30.

Stephens, G. L. (1984). Scattering of plane waves by soft obstacles: Anomalous diffraction theory for circular cylinders. *Appl. Opt.*, **23**, 954–959.

Streekstra, G. J. (1994). The deformation of red bloodcells in a couette flow, Ph.D. thesis, University of Utrecht, The Netherlands.

Streekstra, G. J., Hoekstra, A. G., Evert-Jan Nijhof and Heethaar, M. (1993). Light scattering by red blood cells in ektacytometry: Fraunhofer versus anomalous diffraction. *Appl. Opt.*, **32**, 2266–2272.

Streekstra, G. J., Hoekstra, A. G., Evert-Jan Nijhof and Heethaar, M. (1994). Anomalous diffraction by arbitrarily oriented ellipsoids. *Appl. Opt.*, **33**, 7288–7296.

Sugar, R. L. and Blankenbecler, R (1969). Eikonal expansion. *Phys. Rev.*, **183**, 1387–1396.

Sun, W. and Fu, Q. (1999). Anomalous diffraction theory for arbitrarily oriented hexagonal crystals. *J. Quant. Spectrosc. Rad. Transf.*, **63**, 727–737 (results reproduced with permission from Elsevier).

Sun, W. and Fu, Q. (2001). Anomalous diffraction theory for randomly oriented non-spherical particles: A comparison between original and simplified solutions. *J. Quant. Spectrosc. Rad. Transf.*, **70**, 737–747.

Swift, A. R. (1974). Eikonal expansion as the high-energy limit of the Born series. *Phys. Rev.*, **D9**, 1740–1749.

Tang, C. C. H. (1957). Backscattering from dielectric coated infinite cylindrical obstacle. *J. Appl. Phys.*, **28**, 628–633.

Tobocman, W. and Pauli, M. (1972). Comparison of approximate methods for multiple scattering in high-energy collisions. *Phys. Rev.*, **D5**, 2088–2101.

Turner, L. (1973). Rayleigh–Gans–Born light scattering by ensembles of randomly oriented anisotropic particles. *Appl. Opt.*, **12**, 1085–1090.

Uzunoglu, N. K. and Holt, A. R. (1977). The scattering of electromagnetic radiation from dielectric scatterers. *J. Phys. (UK)*, **10A**, 413–424.

Vaillancourt, R. D., Brown, C. W., Guillard, R. R. L. and Blach, W. M. (2004). Light backscattering properties of marine phytoplankton: relatioships to cell size, chemical composition and taxonomy. *J. Plankton Res.*, **26**, 191–212.

van de Hulst, H. C (1957). *Light Scattering by Small Particles*. John Wiley & Sons, New York.

van Dyck, D. and Coene, W. (1984). The real space method for dynamical electron diffraction calculations in high resolution electron microscopy. *Ultramicroscopy*, **15**, 29–40.

Varshimashvilli, K. V., Demkov, Yu. N. and Ostrovskii, V. N. (1980). *Sov. J. Nucl. Phys.*, **31**, 421.

Videen, G. and Chylek, P. (1998). Anomalous diffraction approximation limits. *Atmos. Res.*, **49**, 77–80.

Volten, H., de Haan, J. F., Hoovenier, J. W., Schreurs Vassen, R. W., Dekker, A. G., Hoogenboom, H. J., Charlton, F. and Warts, R. (1998). Laboratory measurements of angular distributions of light scattered by phytoplankton and silt. *Limnol. Oceanogr.*, **43**, 1180–1197.

von Ignatowsky, W. (1905). Reflexion elektromagnetischer Wellen an einem Draht. *Ann. Physik*, **18**, 495–522.

Voshchinnikov, N. V. and Farafonov, V. G., (2000). Applicability of quasistatic and Rayleigh approximations for spheroidal particles. *Opt. Spectrosc.*, **88**, 71–75.

Wait, J. R. (1955) Scattering of a plane wave from a circular dielectric cylinder at oblique incidence. *Can. J. Phys.*, **33**, 189–195.

Wait, J. R. (1963). Electromagnetic scattering from a radially inhomogeneous sphere. *Appl. Sci. Res.*, **B10**, 441–450.

Wallace, S. J. (1971). Eikonal expansion. *Phys. Rev. Lett.*, **27**, 622–625.

Wallace, S. J. (1973a). Eikonal expansion. *Ann. Phys. (NY)*, **78**, 190–257.

Wallace, S. J. (1973b). High-energy expansions of scattering amplitudes. *Phys. Rev.*, **D8**, 1846–1863.

Walstra, P. (1964). Approximation formulae for light scattering coefficient of dielectric spheres. *Br. J. Appl. Phys.*, **15**, 1545–1551.

Walters, H. R. J. (1984). Perturbative methods in electron – and positron – atom scattering. *Phys. Reports*, **116**, 1 102.

Wang, J. and Hallet, F. R. (1996). Spherical particle size determination by analytical inversion of the UV-visible-NIR extinction spectrum. *Appl. Opt.*, **35**, 193–197.

Weinberg, S. (1962). Eikonal method in magnetohydrodynamics. *Phys. Rev.*, **126**, 1899–1909.

Weiss, K. (1981). Laserstreuexperimente an Einzelteilchen zur Interpretation des optischen Verhaltens interplanetaren Staubes. Dissertation, Ruhr-Universität Bochum (Forschungsbericht BMFT-FB-W 81–047).

Weiss, U. (1974). *Eikonal Expansion of the Scattering Amplitude in ImpactParameter Representation* (Preprint No. DESY 74/9). DESY, Hamburg.

Weiss Wrana, K. (1983). Optical properties of interplanetary dust: Comparison with light scattering of larger meteoritic and terrestrial grains. *Astron. Astrophys.*, **126**, 240–250.

Williams, A. C. (1988). The eikonal approximation without ambiguity in direction. *Ann. Phys. (NY)*, **129**, 22–32.

Wolf, M. (1975). Polarization of light reflected from rough planetary surface. *Appl. Opt.*, **14**, 1395–1405.

Wolf, M. (1980). Theory and application of the polarization–albedo rules. *Icarus*, **44**, 780–792.

Wolf, M. (1981). Computing diffuse reflection from particulate planetary surface with a new function. *Appl. Opt.*, **20**, 2493–2498.

Wriedt, T. (1998). A review of elastic scattering theories. *Part. Part. Syst. Charact.* **15**, 67–74.

Wriedt, T. and Comberg, U. (1998). Comparison of computational scattering methods. *J. Quant. Spectrosc. Rad. Transf.*, **60**, 411–423.

Wyatt, P. J. (1973). Differential light scattering technique for microbiology. In: J. R. Norris and W. B. Ribbons (eds), *Methods in Microbiology* (pp. 183–263). Academic Press, New York.

Xu, M. (2003). Light extinction and absorption by arbitrarily oriented finite circular cylinders by use of geometrical path statistics of rays. *Appl. Opt.*, **42**, 6710–6723.

Xu, M., Lax, M. and Alfano, R. R. (2003). Anomalous diffraction of light with geometrical path statistics of rays and a Gaussian ray approximation. *Opt. Lett.*, **28**, 179–181.

Yang, P., Zhang, Z., Baum, B. A., Huang, H. L. and Hu, Y. (2004). A new look at the anomalous diffraction theory (ADT): Algorithm in cumulative projected-area distribution. *J. Quant. Spectrosc. Radiat. Transf.*, **89**, 421–442 (results reproduced with permission from Elsevier).

Yaroslavsky, A. N., Priezzhev, A. V., Rodriguez, J., Yaroslavsky, I. V. and Battarbee, H. (2002). Optics of blood. In: V. V. Tuchin (ed.), *Handbook of Optical Biomedical Diagnostics*. SPIE Press, Washington, DC.

Yates, A. C. and Tenney, A. (1972a). Multiple-scattering effects in high-energy electron-molecule collision, I: Diatomic molecules. *Phys. Rev.*, **A5**, 2474–2481.

Yates, A. C. and Tenney, A. (1972b). Glauber cross sections for excitation of the 2^1S state of helium by electron impact. *Phys. Rev.*, **A6**, 1451–1456.

Yeh, C. J. (1963). The diffraction of waves by a penetrable ribbon. *J. Math. Phys.*, **4**, 65–71.

Zaneveld, J. R. V. and Kitchen, J. C. (1995). The variation in inherent optical properties of phytoplankton near an absorption peak as determined by various models of cell structure. *J. Geophys. Res.*, **100**, 309–320.

Zege, E. P. and Kokhanovsky, A. A. (1988). Integral characteristics of light scattering by large particles. *Izv. AN SSSR, Fiz. Atm. Okeana*, **24**, 691–700.

Zege, E. P. and Kokhanovsky, A. A. (1989). Approximation of the anomalous diffraction coated spheres. *Izv. Atmosph. Oceanic Phys.*, **25**, 1195–1201.

Zerull, R. H., Giese, R. H. and Weiss Wrana, K. (1977). Scattering measurements of irregular particles vs. Mie theory. *SPIE*, **112**, 191–199.

Zhao, J. Q. and Hu, Y. Q. (2003). Bridging technique for calculating the extinction efficiency of arbitrary shaped particles. *Appl. Opt.*, **42**, 4937–4945.

Zimm, B. H. and Dandlikar, W. B. (1954). Theory of light scattering and refractive index of solutions of large colloidal particles. *J. Phys. Chem.*, **58**, 644–648.

Zocher, H. (1925). Über die optische Anisotropie selektiv absorbierender Stoffe und über mechanische Erzeugung von Anisotropie. *Naturwissenschaften*, **13**, 1015–1021.

Index

Abel inversion 149
absorption
   cross-section 19, 20, 87, 153
   efficiency 19, 30, 41, 67, 124, 157
addition theorems 163
aerosols 1, 125, 142
aggregates 145
albedo 21, 75, 76, 77, 84, 89
algal cells 156
anisotropic cylinder 57
anomalous diffraction approximation 4, 18, 115
  cube 76
  cylinder 51, 81, 85
  disks 87
  ellipsoid 70, 85
  extended 66
  hexagonal column 83, 159
  needles 76
  parallelepiped 77
  plates 76
  sphere 25
  spheroids 70, 72, 78, 79, 85, 86
  statistical interpretation 78
anomalous diffraction theory 115
  simplified 83
asymmetry parameter 21

backscattering 47, 49, 103, 156
bacteria 154

biological tissue 153
biomedical optics 150
birefringent particles 144
blood optics 150
Bohren and Nevitt approximation 114
Born approximation 12
Born series 12

central-incidence approximation 109
circular dichroism 154
coated sphere 66
columnar particles 108
Coulomb potential 13

Debye length 148
deformation index 151
density profiling 148
differential cross-section 20
dust models 146, 148

edge effects 27, 45, 65, 72, 83, 113
eikonal
  amplitude 9, 10, 11, 15, 58
  approximation 17, 18, 24
    corrections 36, 60
    extended 65, 72
    first-order correction 14, 36
    generalized 16, 37
    homogeneous sphere 24

Index

eikonal (*cont.*)
   infinitely long cylinder 51, 59, 95, 96, 82, 109, 137
   modified generalized 37, 38
   optical scattering 17
   potential scattering 5
   second-order correction 14
   series 12
   validity criteria 7, 8, 10, 22, 26, 53
   vector 50, 59
   equation 4
   expansion 14, 15, 36
   picture 38, 43, 44, 133
   wave function 9
eikonal–Born series 14
equiphase approximation 18, 73
Evans–Fournier approximation 18, 110
extinction
   coefficient 136
   cross-section 20
   efficiency 20, 30, 40, 67
   hollow sphere 69
   interstellar 146
   paradox 30
   theorem 20
   sphere 27
   spectrum 138

flow cytometer 118, 128
Fraunhofer diffraction 118, 128
Fresnel reflection coefficients 32

Gaussian ray approximation 80, 154
geometrical optics approximation 4, 114

Hart–Montroll approximation 18, 114
hexagonal columns 76, 83, 84, 159
high-energy approximation 3, 18

integral equation 7
interstellar
   dust 146, 148
   extinction 146

Jobst approximation 18, 93

kaolinite particles
Klett approximation 28

mean value theorem 138, 171
Mie
   particles 24, 138
   solutions 31
modified Rayleigh–Gans–Debye approximation 18, 94

numerical comparisons 13, 41, 63, 118
Nussenzveig–Wiscombe approximation 18, 117

oblique incidence 65
optical fibre 150
optical rotation 154

partial wave expansion 11
Pendorf–Shifrin–Punina approximation 18, 118
Perelman approximation 18, 96
   cylinder 105
   integral representation 100
   modified scalar 104
   scalar 103
polydispersion 21, 88, 154
propagator approximation 8
phase function 20, 40, 156

quasi-static approximation 18, 94

Rayleigh approximation 18
Rayleigh–Gans–Debye approximation 18, 26
red blood cells 150

scattering
  amplitude 6
  coated sphere
    thin 67
    thick 67
  cylinder
    elliptic 66
    normal incidence 51
    oblique incidence 65
  efficiency 20
  ellipsoid 70
  function 21
  sphere
    scalar approximation 23
  spheroid 70

Schrödinger equation 6
Shifrin and Tonna approximation 116
size
   parameter definition 18
   determination 127, 128, 130, 134,135,136
soft-particle definition 18
spectral extinction 18, 137
sum rule 157
surface roughness 146
suspension 136

turbidity 136, 138, 144

visibility 134, 152

Walstra approximation 18,28, 95
WKB approximation 8, 18, 39, 44, 45
   two-wave 18, 40, 49

Zocher's rule 145

Printing: Mercedes-Druck, Berlin
Binding: Stein+Lehmann, Berlin